THE MATHISEN COROLLARY

Mount Everest (left peak) in the Himalayas, the "roof of the world," the highest point on earth. According to the hydroplate theory of Dr. Walt Brown, the cataclysmic events surrounding the global flood produced the tremendous buckling and thickening of the continental plates, and the thickest place on earth is right here. The ten highest peaks on earth are located in the Himalayas.

The creation of the Himalayas acted like a large rock or weight being slapped onto the side of the globe of the earth, and centrifugal force acting on them as the earth spins caused them to want to move towards the equator, initiating a roll in earth's attitude in space. This tremendous displacement would have severely altered the heavens, and would have been noticed by any human survivors of that event, who would probably have studied the skies even more intently after that than they ever did before.

If such heavenly displacements were caused by the events surrounding a cataclysmic flood within human memory, then it would not be surprising to find that the most ancient humans whose records and buildings still exist noticed this phenomenon and recorded it in their monuments and mythologies, and in fact that is exactly what we do find.The pillar atop the chorten in the foreground probably represents the axis of heaven, which features prominently in the mythologies of the world.

THE MATHISEN COROLLARY

CONNECTING A GLOBAL FLOOD
WITH THE MYSTERY OF MANKIND'S ANCIENT PAST

DAVID WARNER MATHISEN

Beowulf Books
Paso Robles, California

Published by Beowulf Books, Paso Robles, CA 93446.

Manufactured in the United States of America.

Illustrations by the author.

Mathisen, David Warner.
The Mathisen Corollary / David Warner Mathisen. -- 1st ed.

p. cm.

1. Mythology. 2. Earth Science. 3. Civilization, Ancient.

ISBN
978-0-615-53562-3

SDG

for my family:

My wife, my sons, my father & mother, and my sister:

Thank you for all of your encouragement, toleration,
and love.

CONTENTS

Introduction

Detective stories, from Sherlock Holmes to Scooby Doo, usually follow a very familiar pattern. The plot almost invariably features an apparent solution that seems to be justified by a superficial reading of the facts. The "experts" or "authorities" are certain of their reading of the situation and confident that their thesis is the most likely explanation. The hero, invariably a marginal figure, arrives on the scene and is able – partially by virtue of that very "outsider" status – to see through the explanation that seems so obvious to everyone else, and to offer a surprisingly different explanation. The outsider (or outsiders, in the case of the Scooby Doo series) startles everyone when he shows how the framework of the conventional thesis is fatally flawed, and in fact explains the mysterious clues that the "experts" or "authorities" mis-read. Often, it turns out that the establishment figures had motives of their own in promoting their thesis, and they are furious at the reversal of their plans, which seemed to be going so well.

There is a growing chorus of voices – generally outsiders or marginalized figures, just as in the detective-story scenario described above – annoying the "authorities" by pointing out larger and larger holes in the consensus explanation of man's origins and distant past. The evidence from archaeological sites around the world – from India to Giza to Europe, Central America, South America and the remote islands of the Pacific – becomes harder and harder to shoehorn into the conventional thesis. The ongoing discovery of new evidence that does not fit into the accepted and interconnected timelines of geology, biology, and anthropology (such as man-made ruins submerged beneath the ocean at great depths) are so upsetting that they are generally ignored by the keepers of the conventional narrative, and those who refuse to politely ignore them as well are generally ridiculed, marginalized, and suppressed.

In fact, the evidence of an advanced, ancient civilization is all around us. Abundant evidence of the level of understanding that it achieved remains in the myths and legends of many cultures, both familiar and unfamiliar, and in the ancient archaeological sites around the world. This book will examine that evidence. It will also bring in evidence from geology and the clues we find all around the world in geological formations and features which indicate that the explanations currently in vogue are incorrect as well. The proper understanding of earth's ancient past and mankind's ancient past are interconnected, as we will see.

It is my belief that the evidence clearly shows that a cataclysmic global flood took place within the timeline of human history, and that human knowledge of this event is encoded in ancient myth and archaeology. This cataclysmic flood caused a change in the orientation of the earth and would have completely altered the view of the heavens for observers on earth. The evidence shows that ancient civilizations were aware of the connection between the flood event and the major change in the celestial sphere, and the very earliest records we have indicate an understanding of the phenomena which is quite remarkable.

This theory completely changes our understanding of mankind's ancient past. It argues that our extremely ancient ancestors – those who lived even before ancient Sumer, Babylon, and dynastic Egypt – possessed knowledge and understanding of astronomical details that conventional historians argue were not understood until the first or second centuries BC, and even then only imprecisely. And yet the preponderance of evidence argues that our ancient anscestors were far more accomplished than we have been led to believe – were far more accomplished than even the most advanced civilizations that came afterwards, such as the Greeks and Romans. We will see that these ancients were not only able to create monuments of great precision using enormous stones and dressed rock but that they understood sophisticated mathematical concepts, the size and shape of the earth, and even how to navigate the oceans and travel between the so-called Old World and New World, which they did frequently.

Such a view of ancient man raises serious questions, such as where these advanced ancients disappeared to, and how their knowledge slowly died away. However, nobody will ever ask these important questions if the very existence of these ancient ancestors is denied and if those who point out the clues that they left behind are ridiculed whenever they do so.

The introduction of geology and earth science into this discussion is an important aspect of this book. The dominant geological model today, the plate tectonics model, while taught as if it were proven fact, may be completely incorrect. Tectonics is really only the most recent incarnation of a larger geological paradigm known as *uniformitarianism,* a framework which argues that almost all the features we find on the earth around us were formed over eons of gradual accumulated changes by forces no different from those going on now.

Uniformitarianism is central to the conventional timeline, because it provides (and actually requires) the hundreds of millions of years that the theory of Darwinian evolution also requires for its assumptions. Thus, the suggestion that catastrophic events could better explain the features we find on earth threatens not only geological models but also biological models. And if these biological models are wrong, then the idea that the most ancient civilization was more advanced than civilizations that followed it becomes more feasible. The conventional timeline of man's slow progression from hunter-gatherer to primitive civilization may in fact be completely incorrect. The geological model and the biological and anthropological models are closely connected.

The evidence we find irrefutably argues that the current geological and anthropological model is wrong. For example, it is easy to see that the ancient pyramids found throughout the world are aligned with celestial phenomena such as the location of important stars on certain dates – often with great precision. Some, such as the Great Pyramid, are also aligned quite precisely with the cardinal directions. The pyramids of Giza are not the only examples of this practice of alignment, but they are certainly the most widely known and heavily studied of the pyramids with celestial alignments.

Descriptions of the Great Pyramid, even by scholars who subscribe to and defend the conventional timelines and anthropological assumptions (of man's slow rise from primitive hunter-gatherer to primitive agricultural villager to eventual civilization and more advanced technological capability), now agree that the pyramid's features point to exact locations of important stars in Orion and Draco.

What is amazing about this piece of evidence is the fact that nobody ever seems to point out how devastating it is for the proponents of plate tectonics. Plate tectonics posits the ongoing movement of huge tectonic plates, upon which the continents and seabeds rest. These plates "drift" atop a semi-molten mantel, causing motion of a little more than an inch per year.

If the plate carrying the pyramids at Giza (or any of the other pyramids located around the world, which remain perfectly aligned with the stars and with the cardinal directions) has been drifting by as much as 30mm or a little over an inch per year, then even using the accepted dates of the most conservative conventional scholars which place the pyramids and the Sphinx in the epoch of 2600 BC, we would expect the Earth to have carried them well over 4,600 inches (383 feet) from their original positions, eliminating any trace of their original alignments. And yet the alignments remain as precise today as when they were built. This fact argues strongly against the tectonic theory, and yet the divide between geology and anthropology appears to have prevented scholars on either side from making the connection.

Note that "drift" of even a few inches could destroy the precise star alignments which are present in the pyramids, to say nothing of hundreds of feet. Also note that it is not only the pyramids of Egypt which contain star alignments, but also those of Central America, along the "ring of fire" (where plate movement should be quite significant, according to the tectonic model). And while those pyramids are supposedly more recent than those at Giza, most scholars believe that Stonehenge is far older – dating to 3000 BC or earlier. Yet Stonehenge still retains its own precise stellar and solar alignments, although it should have drifted even farther than Giza. The fact that these star alignments are still present in all these varied places indicates that we should be open to other models besides tectonics for explaining the geological features of our planet.

These examples indicate the importance of examining the anthropological frame or paradigm in light of the geological frame or paradigm, and the geological in light of the anthropological. Such a unified approach is sadly lacking in the examination of many of the amazing clues on our planet's surface and of the equally amazing clues to be found in both the ancient architectural monuments and the recorded mythologies which provide us with priceless windows into the ancient past.

Another central tenet of conventional geology is the belief that the geologic strata were laid down over hundreds of millions of years. Therefore, when archaeologists find evidence of human artifacts or human remains in strata assumed to be hundreds of millions of years old, some argue that *homo sapiens* has in fact been around for hundreds of millions of years, while others throw out the artifacts or remains as impossible and therefore an obvious hoax. However, if in fact the underlying geological model which proposes that the strata were laid down in successive ages is incorrect, then neither of these two anthropological conclusions remains necessary.

In fact, there are solid scientific reasons for believing that the strata were not laid down over successive periods of hundreds of millions of years. The main evidence reinforcing that faulty model rests upon the consistent presence of various types of fossils within particular strata. The evolutionary assumptions requiring millions of years of mutation and natural selection between these various fossil species leads to (and requires) the vast periods of time between strata.

In his examination of geological and astronomical evidence, West Point graduate and former professor Dr. Walt Brown has pointed out several flaws with the reigning geological models (including the tectonic theory and the assumptions of great age between strata) and offers a contending model. Dr. Brown's model, which he calls the hydroplate theory, explains numerous geological features far better than the conventional theories, and his arguments rest, not on an *a priori* commitment to a specific viewpoint, but on an examination of the primary conventional explanations for various phenomena and a comparison of their relative merits at each point, as well as a frank assessment of the ability of his theory to explain the same points (when his theory has a weakness in accounting for a particular detail, he admits it).

It is my contention that, in addition to the extensive geological and astronomical evidence examined in Dr. Brown's writings, numerous mysteries of human archaeology and culture (many of which have been documented by writers offering analysis outside of the mainstream of academic consensus) are best explained within the framework of Brown's hydroplate theory, and form important additional supporting evidence for his theory versus conventional explanations based on tectonics and other uniformitarian frames.

Erroneous models can lead to big problems. Some scientists, for example, have argued that the misinterpretation of ambiguous data from a few dietary studies in the 1950s led to erroneous dietary campaigns that have actually increased the incidence of heart disease, stroke, and diabetes. The question of who is right in this case carries important consequences for everyone.

Geological models have consequences as well. The tectonic model underlies our conventional earthquake explanation and prediction efforts. Incorrect models could lead to unnecessary loss of human life, while more accurate models might help prevent loss of life. Correct geological models are also extremely important for mining and the pursuit of natural gas and oil. Conflicting models exist regarding the origin of these fuels, which lead to different conclusions about how much of them remain for future extraction. If a model is flawed, it will lead to less accurate predictions.

Geological models also impact biological models. If the uniformitarian assumptions are incorrect, this fact will have devastating ramifications for Darwinist models of evolution. For example, if the Rocky Mountains were pushed up by powerful and unprecedented forces only thousands of years ago, as opposed to millions of years ago, it becomes far more difficult to argue that species such as beavers or bighorn sheep evolved their highly complex behaviors and adaptations to those environments over millions of years. A Darwinian model often informs advice about how we should live, what we should eat, what behaviors are healthy and what behaviors are not. If these models are incorrect, they can have profound consequences.

Finally, what we might call "anthropological models" – narratives about man's ancient past – have important consequences as well. If ancient civilizations crossed the oceans, comprehended the size and shape of the Earth, and achieved levels of mathematical and astronomical understanding several hundred or even thousands of years before known civilizations could do so, it is important to know for several reasons. We might want to take a closer look at any messages those ancient people may have succeeded in preserving through the millennia. In fact, as we will see, they were able to do so, through the medium of mythology, as well as through their enduring monuments and artifacts.

Operating from a flawed anthropological model can lead to complacency about our position in history. The conventional narrative is comforting in many ways: we can look back from our privileged position in history and contemplate the long, slow, painful progress that brought mankind to this point in time. We are tempted to believe that progress is the natural state of affairs, and that it is an inevitable feature of human experience.

But this paradigm for examining human history may be completely incorrect. It may be instead that man possessed great knowledge in the distant past and then lost it. If so, that fact raises a completely different set of questions that we should be asking ourselves than the questions we might tend to ask ourselves today, based on our incorrect assumptions.

For all these reasons, the examination of this subject is no mere curiosity. Of course, our curiosity can play a role as well – who wouldn't want to know more about the amazing mysteries of mankind's ancient past, many of which have been suppressed orhidden by those who are invested in the conventional explanation? Who doesn't enjoy a good mystery story, and who would rather be satisfied with the storyline that the "authorities" are trying to push, even if that storyline is not what really happened? While no human book has all the answers, and all will contain some mistakes, it is hoped that the examination which follows will stimulate more detectives to examine the fascinating case of mankind's ancient past.

The Hydroplate Theory of Dr. Walt Brown

There is extensive evidence that the conventional explanation of history is fatally flawed, but that its proponents are strongly biased towards the dominant theories of anthropology and geology because these theories are critical reinforcements for the Darwinian theory of evolution in biology.

We will see that not only does the conventional self-reinforcing framework of biology-anthropology-geology contain fatal flaws, but that an alternative catastrophic geological framework fits the existing evidence much better. Other challengers to the conventional explanations have put forward catastrophic explanations involving near-collisions from Venus, or the "slippage of the earth's crust like the skin of an orange that somehow came loose" (in the "Earth-crust displacement theory" of Charles Hapgood described in his 1966 *Maps of the Ancient Sea-Kings,* page 187, and also in Graham Hancock's 1996 *Fingerprints of the Gods,* page 10) but these speculative explanations do not solve many of the geological problems presented by the physical evidence in the world around us. In contrast, the theory put forward by West Point graduate Dr. Walt Brown is backed by rigorous scientific explanations for the extensive evidence in the geological record, which he documents in his text, *In the Beginning: Compelling Evidence for Creation and the Flood,* currently in its eighth edition.

While the evidence Dr. Brown has mustered from the geology of the earth (as well as from certain phenomena in the solar system, including the asteroid belt, and the characteristics of comets, and even the presence of water and ice on the moon and Mars) makes his theory worthy of consideration, the thesis of this book is that there is further supporting evidence to be found in the clues left to us from ancient civilizations. Specifically, there are clues in the mythologies of ancient cultures that are a kind of "literary artifacts" every bit as important as the physical archaeological artifacts remaining from the ancients. Beyond these mythological clues, there are the physical archaeological artifacts themselves, and together these clues from myth and from ancient archaeology provide additional strong support that the current uniformitarian models built upon tectonics are incorrect, and that the earth did experience a catastrophic event similar to that described by Dr. Brown.

Before we begin to examine the evidence from myth and from archaeology, a brief overview of Dr. Brown's hydroplate theory is in order. We will return to it in greater detail later, after some of the evidence from myth and archaeology has been discussed.

In sum, the hydroplate theory posits a global flood originating with water that was trapped beneath the earth's surface (while this may sound like an astonishing starting point, Dr. Brown provides extensive evidence for just such a catastrophe, and amazingly, world mythology not only tells of a global flood but often insists that the waters came from under the earth).

As difficult as this starting point may be for those who have been told since grade school that plate tectonics explains all of the features we see in the earth around us today, it is important to keep in mind that the theory of plate tectonics itself was ridiculed and viewed as a "fringe theory" from the time of its invention in the early decades of the twentieth century until its adoption by the mainstream in the late 1960s. Its acceptance came as a result of the observations of the deep ocean floor which were not made possible until the 1950s and 1960s. Alfred Wegener (1880 – 1930) who is credited with the championing the theory of tectonics and continental drift beginning in 1912 was fiercely ridiculed for it and did not live to see its later acceptance by the academic world.

As a child growing up in the early 1970s, I was taught the theory of plate tectonics by my forward-thinking teachers, but at that time it was still a fairly novel theory and children were by no means exposed to it everywhere. Today, the theory of plate tectonics is taught as though it is the only possible explanation, and anyone who questions it is subject to ridicule. This is ironic, since its proponents were subjected to the same ridicule for many decades, during which the proponents of the then-reigning paradigm were absolutely certain that they were the keepers of the true explanation, and the upstarts who dared to challenge them were vehemently opposed by the highly respected and heavily published defenders of the old order, who held all the posts of importance in the geological community.

This history as a fringe theory does not, of course, provide actual evidence that tectonics is not correct, but it should offer cause for some humility from its current proponents, who now find themselves in the position that their theory's opponents once occupied.

Dr. Brown offers several pieces of evidence for the position that the current tectonic paradigm, while an improvement over previous theories, is flawed and in need of major surgery. First, the features of the deep ocean floors, which were so persuasive in the eventual adoption of the theory of plate tectonics, actually argue for catastrophic explanations and not for the gradualist mechanisms of tectonics.

The original opponents of tectonics, most of whom could be categorized as "fixists" who believed that the continents never moved from their current location, largely argued that the ocean floor was too solid and too strong to allow continents to wander about the way Wegener and his followers proposed. When technological advances during and after the Second World War enabled scientists to map the ocean's bottoms, and to take readings of the magnetic fields over various submarine features, the ridges and chasms that were discovered destroyed the main objection of the fixists, and seemed to confirm the predictions that Wegener had made decades earlier (see the discussion of "The Continental Drift Debate" by Henry Frankel, in Hugo Tristram Engelhardt and Arthur L. Caplan's *Scientific Controversies: Case Solutions in the Resolution and Closure of Disputes in Science and Technology*, 203-248).

Dr. Brown points out, however, that many of the seafloor features that at first led to the widespread adoption of the tectonic theory actually argue against it. For instance, the Mid-Atlantic Ridge, an extremely important feature running north-south through the center of the Atlantic Ocean, and which actually runs for over 46,000 miles around the earth, intersecting itself "in a Y-shaped junction beneath the Indian Ocean," contains numerous "fracture zones" that are offset in ways which make it extremely difficult to explain how separating, shifting, or colliding tectonic plates could have caused them. The fracture zones overlap, curve, and intersect in ways that cannot be explained by the movement of plates (Brown 87-90).

Similarly, Brown points out that the deep ocean trenches that ring much of the boundaries of the Pacific basin "are often shaped like long arcs that connect at sharp cusps" (121). The plate tectonic theory has a very difficult time explaining deep ocean trenches shaped in great sweeping arcs that join other arc-shaped trenches at a sharp point or cusp. The conventional tectonic explanation for the origin of such deep trenches involves the subduction of one massive plate beneath another, which could perhaps explain a long straight trench (although Brown provides convincing arguments from physics that the pressure required to subduct a miles-thick plate below another miles-thick plate is beyond even the forces that proponents of the tectonic theory propose for these plates). However, the tectonic theory would have tremendous difficulty explaining a highly-curved trench, and especially a curved trench next to another curved trench that come together in sharp cusps, as illustrated on the following page.

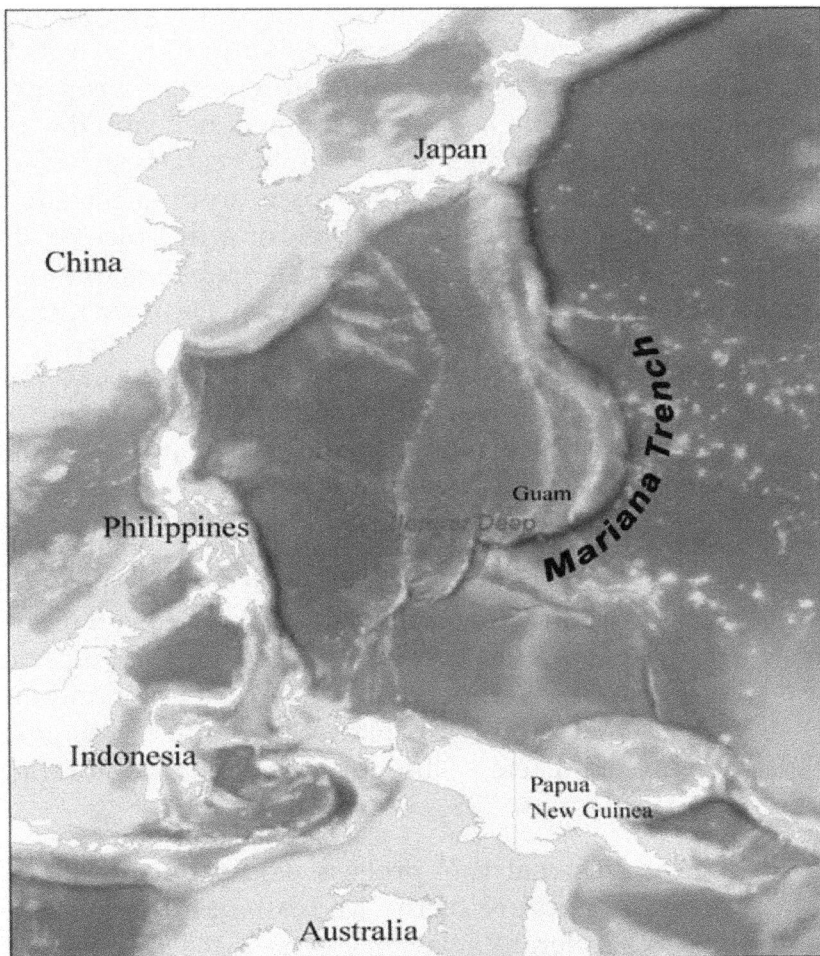

The characteristic arc-and-cusp pattern of deep ocean trenches in the Pacific is illustrated above, and can be seen on any map of the Pacific ocean floor. Notice the Mariana Trench and the cusp at the southwest end of the trench, west of Challenger Deep.

Dr. Walt Brown observes that this arc-and-cusp pattern is very difficult to explain using the tectonic theory.

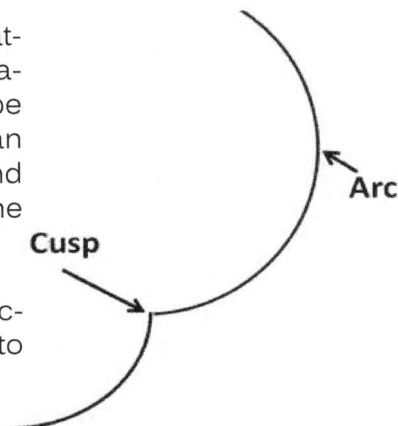

The hydroplate theory of Dr. Brown argues instead that the globe-encircling scar known as the Mid-Atlantic Ridge in its Atlantic segment is actually the remnant of an enormous rupture which released water which had been trapped beneath the pre-flood globe at great pressure, causing a global cataclysm. This water blasted upwards with tremendous force, eroding the edges of the continental plates above the rupture, for a distance of about 400 miles. The physics associated with this globe-encircling rupture caused a period of rapid continental drift. Dr. Brown explains:

> Eventually, the width [of this rupture] was so great, and so much of the surface weight had been removed, that the compressed rock beneath the exposed floor of the subterranean chamber sprang upward.
>
> As the Mid-Atlantic Ridge began to rise, creating slopes on either side, the granite plates (which we will call hydroplates) started to slide downhill. This removed even more weight from what was to become the floor of the Atlantic Ocean. As weight was removed, the floor rose faster and the slopes increased, so the hydroplates accelerated, removing even more weight, etc. The entire Atlantic floor rapidly rose almost 10 miles. 102.

Dr. Brown argues that this process would have taken place all along the entire rupture of the Mid-Atlantic Ridge, causing the continents to accelerate away from the ridge in both directions, lubricated by the water "still escaping from beneath them" (102). However, as the water lubricant beneath each plate was depleted, or the plates ran into something on their forward edge, "each massive hyrdroplate decelerated" and experienced "a gigantic compression event – buckling, crushing, and thickening each plate" (102).

The rapid compression caused mountains to form and overthrusts to occur, generally along north-south axes perpendicular to the motion of the plates as they slid away from the original rupture. Magma created from the tremendous heat along the leading edges of the sliding plates also spurted out in many places, creating volcanoes and other formations found today in areas that experienced the most friction. The thickening that took place when the plates ground to a halt caused the water to run off to the ocean basins, which were far lower than they are today. Later, the weight of the continents caused them to sink lower and force the rise of the ocean bottom and the ocean level to its present height, as well as causing the rise of the extensive plateau areas found adjacent to the earth's mightiest mountain ranges. Previous to this sinking, the ocean levels were lower (and warmer, due to the energy released during the cataclysm), and the continents relatively higher, creating wide land bridges between continents, and also setting the conditions for an ice age (warm oceans generated more clouds and precipitation, and high continents meant that the precipitation was often snow and ice).

The upward movement of the floor of what today is the Atlantic Ocean following the initial rupture, particularly the upward movement of the enormous submarine mountain range that is the Mid-Atlantic Ridge, caused an opposite movement on the other side of the globe, sucking the floor of what is today the Pacific Ocean basin towards the earth's center. This force created the arc-and-cusp trench patterns and the many amazing features peculiar to the Pacific Ocean, including its "ring of fire" of volcanic activity around its borders, and the periodic settling of the continents towards this vast abyss continues to be felt as earthquakes around its perimeter (and below its surface) to this day.

According to the theory, when the water was running off the continents into the oceans (which initially had lower levels than they do today, due to a process explained later), the massive flows often carved enormous canyons in the continental shelves, now submerged in water after the continents sank and the oceans rose centuries later. These massive submarine canyons can still be seen today (often beginning at the mouths of mighty rivers where they enter the sea, such as the Amazon or the Ganges or the Hudson) but are now covered by water since the ocean levels have risen significantly. Brown notes that it is very difficult to explain these massive canyons in the sides of the continental shelves under conventional theories (92). Carving such canyons under water is very difficult to explain, but runoff from the continents before the oceans rose to their current levels would explain these enormous canyons quite satisfactorily.

Also according to this theory, while most of the water flowed off of the thickening continents, initially some of it was trapped in various geographic areas and uplifted into high-altitude lakes with tremendous potential energy. In some places, subsequent precipitation and runoff caused these lakes to increase in size and some burst their boundaries, causing violent and massive flows of cascading water which carved distinctive geologic features as they sought the ocean, as well as creating unique formations within the former lake boundaries as the water level fell rapidly. Brown compares this explanation for the formation of the great canyons found next to major plateaus – such as the Grand Canyon next to the Colorado Plateau or the Jhelum River canyon emanating from the Kashmir Basin in the Tibetan Plateau – to the explanations of uniformitarian theories, and finds that the hydroplate theory explains numerous geological features that the uniformitarian theories cannot explain.

In other places, where no violent breaching of uplifted lakes took place, such as in regions where there was little precipitation, the trapped water slowly evaporated over centuries, leaving dry lake-beds such as are found around Lake Titicaca in the region of Peru's border with Bolivia near La Paz, or the Great Salt Lake in Utah.

In some places, Brown explains, the path of the globe-encircling rupture was later over-ridden by a sliding continent, for example along the forward-sliding edge of the North American continent. Today, that rupture line is beneath the portion of North America stretching from California to Alaska, but its path can be traced by drawing a line along the ancient volcanoes that also stretch from Southern California to Alaska (some of which, such as Mt. St. Helens, are still active in our time; Brown 113, note 31).

Dr. Brown labels the phases of the cataclysmic events described above as the rupture phase, the flood phase, the drift phase (culminating in a massive compression event), and the recovery phase, the last of which continues today. He demonstrates how this series of events provides a better explanation for numerous features in the world which are difficult or impossible to explain by conventional uniformitarian theory. A key point that Dr. Brown argues is that such a flood would, by the principles of liquefaction, create the sorting and layering of the strata that are found around the world, as well as other liquefaction features such as Ayers Rock in Australia.

The mistaken belief that the strata were laid down over hundreds of millions of years forms a key argument in the conventional paradigm for both biology and geology. However, the uniformitarian processes of the conventional theories have a very difficult time explaining the actual details of the strata (such as the presence of polystrate fossils which pierce through several strata, or the fact that the strata generally have sharp and well-defined boundaries, which is not what we would expect if each layer was exposed to erosion for millions of years before the deposition of the next layer). The hydroplate theory provides a much better explanation for the actual details that are found in the strata, as Brown details in his text.

The conventional framework depends quite heavily upon the mistaken identification of the strata as indicators of great ages of earth's past. In fact, conventional biology and geology use a very circular argument based on these strata. The biologists argue that Darwinian evolution took place over great ages as evidenced by fossils found in various layers, which are held to have been laid down during certain successive periods of hundreds of millions of years. However, if asked how we know when those ancient strata were laid down, the age of the fossils found in those layers will often be used as an argument for their age. In other words, the fossils must be ancient because of the age of the strata in which they are found, and the strata can be known to be of that age because of the age of the fossils which are found therein!

Brown's theory explains the origin of the strata found around the world quite differently. The initial rupture would have eroded tons of sediments, making the waters that covered the earth during the flood extremely sediment rich. The action of powerful waves would create the conditions for a phenomenon known as liquefaction, which takes place even today under certain conditions, including when powerful waves pass over the sandy ocean floor. Liquefaction involves sediments falling down through water (and water rising upwards past sediments), which would have sorted the sediments by type. Dr. Brown argues that the flood phase would have tended to sort both the various types of particles and the various types of buried organisms.

Such an explanation would explain the origin of all of the world's fossils. Note that fossils do not form under normal circumstances (a powerful clue that uniformitarian explanations of geology may be flawed). Animals and plants which die at sea or on land do not

image: Wikimedia Commons http://upload.wikimedia.org/wikipedia/commons/2/25/Folding_Gasterntal.jpg

Swiss mountainside showing a graceful curving pattern in the strata. The theory of Dr. Brown argues that the strata found around the world were laid down during the flood phase by the process known as liquifaction and were still soft during the compression event at the end of the continental drift phase which was initiated by the rupture and the flood.

usually leave any remains which can form a fossil. The sudden onrush of sediment-infused water, however, could have buried plants and animals and sealed them off from the bacteria which under normal circumstances would break down even the bones to nothing.

This explanation would also explain fossils which have been found penetrating multiple strata (polystrate fossils, such as fossil tree trunks going through dozens of layers), or the presence of the anomalous quartzite block which can be found above the "Precambrian-Cambrian" interface in the layers of the Grand

Canyon, in the middle of a completely different sediment and layers above the quartzite layer that is generally considered to be Precambrian (145). Dr. Brown also observes that the many places on earth which display severe buckling of an entire series of strata are very difficult to explain if those strata were laid down over several hundred million years, hardening each time (see the illustration from Gasteretal, Switzerland). Instead, he says, "the layers had to have been soft, like wet sand, at the time of compression" (94). He also notes that strata seldom show signs of millions of years of erosion between them, but rather demonstrate crisp clear borders between each layer of the strata (30).

Several of the features of Brown's theory, if correct, would also help to explain the many anomalous clues left to us by ancient civilizations in their mythologies and their ancient architecture. Although Brown does not generally go in this direction, it is the thesis of this book that these pieces of evidence from myth and ancient archaeology not only support many aspects of Brown's theory but also find their most compelling explanation when fitted to the cataclysmic events Brown describes, and the conditions which would have prevailed upon the world in the aftermath of such an event.

The few places in his text that Brown addresses anthropological matters can give a hint at the value of such application of his theory. He argues that the lower sea levels that prevailed for some centuries would have provided many migration paths for men and animals to populate today's continents and islands, and notes that the oral history of the Hopi as told to Frank Waters in the *Book of the Hopi* recounts:

> After a gigantic flood, their ancestors used many family-sized rafts and "island hopped" for many years north and east to the Americas. The steep coast line [today's continental slope] forced them along the coast until they could land. Rising water later drowned the chain of islands. 255 (explanatory material in brackets is Walt Brown's observation in his original text).

This application of the hydroplate theory to the evidence from humanity's past is tantalizing. What other mysteries could the hydroplate theory explain if it were applied to anthropological questions as well as to the largely geological questions Brown treats in his work?

We shall examine many of them in this book. As a prelude, it would certainly provide an explanation for the sunken archaeological remains found in the Maldives, as well as those found along the coast of India, and off some of the islands in Japan, ruins which Graham Hancock discusses in his book *Underworld*. Other sunken archaeological remains have also been attested off the coast of Cuba and in other parts of the Western Hemisphere as well.

The hydroplate theory would also clear up a mystery which I have never read anyone discussing, the mystery of how – if the tectonic theory is correct – the ancient monuments of the world such as the Great Pyramid, the Sphynx, the Sun Gate of Tiahuanaco, and the numerous pyramids of Central America are still aligned with astonishing precision to the rising of the sun and the angles of specific stars. Brown's theory argues that the theory of slow and ongoing continental drift is incorrect. There was a period of massive drift, which left geological evidence that has been wrongly interpreted to be evidence of ongoing drift. Instead, there is ongoing *shifting*, rather than drifting. Brown explains shifting by introducing the metaphor of a large cargo truck hastily packed with heavy boxes stacked on top of one another. Occasionally, the motion of the truck will cause the boxes in the back to shift to find a position of lower potential energy (Brown 121). Occasional shifting may explain some earthquakes, but unlike the mechanism of drifting which the tectonic theory relies upon, it does not have difficulty explaining why extremely ancient monuments are still fairly well aligned with their intended astronomical targets.

Finally, the hydroplate theory proposes, as we will discuss in greater detail in chapter seven ("Precessional Numbers"), that the orientation of the earth – and hence the orientation of the sky – was ominously altered when the massive Eurasian plate thickened and the Himalayas were pushed upwards. According to the hydroplate theory (which, like other catastrophic theories, does away with the need for long ages of slow change that the Darwinists find so comforting as supporting arguments for their theories), these events took place during human history. If so, then this theory would explain the world-wide mythological evidence linking the chopping down of the celestial axis with the initiation of a massive flood, and explain the ancient understanding of precession which goes against all the timelines held by the keepers of the conventional paradigm. Conversely, the worldwide myths would provide supporting evidence that this cataclysmic event took place within human history, just as Walt Brown's theory says that it did.

Ruamahanga Woman (stay away from me)

In October of 2004, a young local of New Zealand's rural Wairarapa Valley was taking a walk along the Ruamahanga River with his two dogs. The sixteen-year-old was surprised to find the crown of a human skull protruding from the gravelly river sand that had recently been under the flood waters not uncommon for the river, particularly at that time of year.

If New Zealand shares many geographic features with California, as some have observed, in spite of the fact that New Zealand is surrounded on all sides by ocean and cut in half by the water of the Cook Straight, the Wairarapa in shape and size might in many ways resemble the northern reaches of California's great Central Valley, although abbreviated by the presence of the strait that divides the northern and southern islands of Aotearoa.

The Ruamahanga is a broad, meandering river that plows its way through the very center of the wide Wairarapa, a valley of farms, dairies, vineyards, and sheep pastures stretching between the Rimutaka Ranges on the west and the Tararua and Aorangi Ranges to the north and east. Almost exactly 100 miles in length, it is fed by major tributaries including the Waiohine, Tauhereniku, Waipoua, Waingawa, Kopuaranga, Whangaehu, Tauweru, and Huangarua Rivers.

Monthly flow in the Ruamahanga River in its southern reaches, near Pukio and Featherston where the skull was discovered, typically varies from a low of about 40 cubic meters per second in January and February, to around 120 cubic meters per second in August, September and October, with a high of up to 140 cubic meters per second in July (the heart of winter in the Southern Hemisphere).

October is typically the wettest month in the upper reaches of the Tararua Ranges, where the Ruamahanga has its origins, and that spring (October being in the spring in the Southern Hemisphere) the river had run high for several days near Sam Tobin's family farm in Pukio. When the youth told the authorities about the human skull he had found, the subsequent investigation ultimately led to the shocking conclusion by forensic scientists that the skull had belonged to a European woman, aged 40 to 45 years, who had died around the year 1654, plus or minus another thirty-five years (a bracket from 1619 to 1689).

The presence of a European woman in New Zealand in that period of time causes major problems for conventional historical timelines, which credit the first European contact in New Zealand to Dutch explorer Abel Tasman (1603 – 1659), who in 1642 sailed from the west (in fact, from Tasmania) and skirted along the western coast of the northern portion of the South Island and the southern portion of the North Island before leaving New Zealand and proceeding north to Tonga and Fiji. No other European ship is known to have returned to New Zealand for over a hundred years.

If the analysis of the skull's ethnic characteristics and date of origin is correct, it creates a very difficult situation for those who would explain its presence in the quiet Wairarapa along the banks of the Ruamahanga. There were no women recorded as being part of Tasman's expedition, and indeed it would have been remarkable had there been any women on board such an expedition in the first half of the 1600s.

Further, Tasman's actual physical contact onshore was limited, due to the incident which took place the first morning that they determined to "get ashore and find a good harbor" (J.E. Heeres, *English Translation of the Journal of Abel Janszoon Tasman*). Tasman's two ships, the *Heemskerk* and the *Zeehan*, anchored about a half mile off the coast of what is now Golden Bay, in 13 to 15 fathoms of water, were preparing to go ashore, while the inhabitants of the islands (the Maoris) headed out into the waters around the anchored ships with first one war-canoe (or *waka*) and then seven more. According to Tasman's journal (translated into English and available online), one of the wakas had seventeen "able-bodied men in her" and another had thirteen.

As the cock-boat of the *Zeehan* headed towards the *Heemskerk* with the quartermaster, the men on one of the wakas began paddling furiously in unison and rammed the cock-boat with terrific force. One of them struck the quartermaster in the neck with a long Maori paddle so violently that he fell overboard, after which the other men in the waka leapt into the cock-boat and began to bludgeon the Dutch sailors. Those who could get away leapt into the water and swam to the safety of the *Zeehan*, where they were hauled aboard. Three Dutchmen were killed and a fourth mortally wounded.

After that, Tasman and his men concluded that "we could not hope to enter into any friendly relations with these people, or to be able to get water or refreshments here." They proceeded up the coast and attempted twice to put in for water, sending the cock-boat of the *Zeehan* and the pinnace of the *Heemskerk* and arming the rowers with pikes, muskets and side-arms, but found that both places they tried were too difficult to approach due to large surf and dangerous rocks, and also due to the presence of Maori warriors whom they had seen standing as lookouts along the high ground all along the Dutchmen's voyage northwards.

They decided to get fresh water later and proceeded north to Tonga, where they found the inhabitants less violent.

This visit is the only known contact with New Zealand by any European vessels until the first voyage of James Cook (1728 – 1799), who reached New Zealand from the east (sailing from Tahiti) in 1769 and mapped its complete coastlines. One hundred and twenty-seven years had passed since the abortive encounter by the expedition of Tasman.

From the record of Tasman's journal, even had his expedition had women or a woman on board (which it did not), it would have been extremely unlikely that a woman would have been included in the heavily-armed watering party that was sent towards shore after proceeding up the coast of the North Island from Murder-er's Bay (as Tasman called the magnificent bay at the north end of the South Island, now known as Golden Bay, where his four men were killed).

Even assuming that somehow an unrecorded woman was on the Tasman expedition, and that she had somehow been selected for the arduous landing in heavy surf to fill the heavy water-casks in enemy territory (an unlikely assumption), Abel Tasman's journal records that the watering party never made landfall due to the concerns of the pilot-major, who feared that an attempt to land on the rocky shore would wreck the boats or result in the smashing-in of the water casks.

However, since Tasman's expedition was the only known European expedition to reach New Zealand in the entire century of the 1600s, there remain few other possibilities within the conventional framework of human origins and history.

Perhaps an unrecorded ship from another European country was blown so far off course that it reached New Zealand, and that unrecorded ship somehow had a European woman or European women aboard, in spite of the fact that no women are known to have been taken on any such expeditions. However, no expeditions exist which can be offered as possibilities for such an "off-course" New Zealand contact.

The only other explanation for a woman of European ethnicity being present on the North Island of Aotearoa in the century of the 1600s is that Europeans were somehow present there prior to the voyage of Tasman. Since no New Zealand inhabitants of European ethnicity prior to Tasman are known to recorded history, this explanation raises the specter of a lost history, and the crossing of the oceans by members of the "Old World" long before the keepers of the current storyline will allow.

Thus the skull of the Ruamahanga Woman is a very important data point for any rigorous analysis of the question of mankind's ancient past, and ignoring a data point of this significance can lead to dangerously flawed conclusions.

In the world of investment research, prior to committing millions of dollars, tens of millions of dollars, or hundreds of millions of dollars of capital to a company (through the purchase of stock issued by that company, or through the purchase of that company's bonds), thorough research and analysis and the consideration of every available data point is imperative. Ignoring a piece of information about that company – especially one which has been written about in respectable publications, and especially one that could have serious implications, such as a report of unethical behavior in the CEO's past, or evidence that the company's business model is vulnerable to a rising alternative means of delivering that company's good or service to customers – could very well lead to a flawed investment thesis and to serious losses.

Analysts have an expression for the kind of thorough investigation and analysis that is required in order to mitigate against a mistaken investment thesis, a level of analysis they call "due diligence." The word "due" means "deserving" – the level of research and analysis that a subject of that importance deserves. Thus the word "due" in English also implies payment – the wages which are "due" a worker, which he has earned by his labor. For those whose entire profession is anthropology or ancient human history, to ignore the information provided by the skull of the Ruamahanga Woman when formulating a thesis about the history of mankind would be as egregious as it would be for an analyst at a major investment firm whose entire profession is investment research to ignore a glaring piece of evidence that provides a window into the leadership or business model of the company he is researching, and to ignore it because it contradicts his personal investment thesis on the company.

Similarly, in the world of military tactics and military intelligence, when officers of a combat unit are formulating their plan for an operation, every available piece of relevant information about the disposition and possible actions of the enemy must be analyzed carefully. Unlike the situation in investment analysis, in which millions of dollars might be on the line, in a military operation men's very lives are on the line. If there are reports available from the scouts, for example, of enemy activity in a certain area, and the commander insists that the enemy could not possibly be coming from that direction (because that would conflict with the commander's favored theory of what the enemy was planning to do), such negligence could very well have tragic consequences.

However, in the academic world where theories of mankind's ancient history are cultivated, the immediate consequences of faulty analysis or for holding on to an incorrect thesis do not generally entail the loss of millions of dollars, as it does in the investment world, or the loss of the lives of those the theorists know well (or loss of the lives of the theorists themselves), as it does in the military world. On the contrary, in the academic world, challenging a well-entrenched theory can lead to the loss of money, status, even livelihood.

From the perspective of a military tactician or an investment analyst, the implications of the Ruamahanga Woman are enormous. If a data point of its size were to drop into the lap of a military intelligence officer while participating in the military decision-making process before a battle, he would have to immediately and urgently seek to corroborate whether it was creditable, and if it led the leaders of that unit to throw out the entire previous plan and switch to a new one, so be it.

If while conducting due diligence before committing capital to a company an investment manager were to stumble across something as damaging to the investment thesis as the existence of the Ruamahanga Woman is damaging to the conventional historical thesis, that manager would immediately put his plans of committing capital to that company on hold while he looked into it further, and if it appeared at all possible that the information it was telling him could be true, he would change his investment thesis and even abandon plans to commit capital to that investment without any hesitation.

Naturally, two lines of investigation for "due diligence" would be the evidence supporting the conclusion that Ruamahanga Woman was in fact a female of European origin, and the evidence supporting the conclusion that the owner of the skull ceased living between 1619 and 1689.

Certainly, skulls themselves by their physical features can indicate both the sex and ethnic characteristics of their owners. The scientific fields of forensic anthropology and bioarchaeology study the morphology of human skeletal remains to enable analysis of sex, ethnic category, pathology (the impact of disease and sickness), and the cause, manner, and mechanism of death from those most physically durable members of the human body, the bones.

There are morphologies in the skull which can be measured to distinguish between the skull of a male and the skull of a female. While the differences in the pelvic bone are the most obvious indicators of whether a skeleton belonged to a man or a woman, there are differences in the skull as well, including differences in the supraorbital margin (the lower edge of the frontal bone, a piece of the skull which makes up the forehead area and terminates at its lower margin in the upper arches of the eye sockets or orbits, the actual bony edge of which will be more rounded in a male and more sharp like a knife-blade in a female), the glabella (another portion at the lower front edge of the frontal bone, between the two arches of the eye sockets, and tending to be flatter in the skull of a woman), the gonial angle (or *angulus mandibulae*, the angle formed by the lower jawbone or mandible, which has one branch or *ramus* ascending towards the ear on either side and a more horizontal portion which runs around the chin, and the angle between these on either side, below the ear, is known as the gonial angle and is typically closer to a right angle in the skull of men and a more obtuse angle in the skulls of women – the skull of the Ruamahanga Woman, however, lacked a mandible when discovered), the shape and curve of the palate (also not present in the remains of the skull of Ruamahanga Woman), the size of the mastoid process (the portion of the skull where it dips down below and behind the ear, tending to be larger in males), and at places where muscles attach, such as the occipital bone (males tending to have more pronounced muscle attachments than females).

In the case of the skull of the Ruamahanga Woman, forensic anthropologist Dr. Robin Watt, of Wellington, NZ, examined it in 2004 and stated that it was probably of a woman of European descent. Dr. Watt, who has practiced for over thirty years, has examined and analyzed more early prehistoric Maori skeletal material than anyone else, according to an extensive 2009 article in *New Zealand Geographic* authored by Vaughan Yarwood. Dr. Watt's determination that the skull probably belonged to a woman and conformed to a European ethnic pattern is significant, because he himself does not admit to the possibility that the skull could point to the presence of people of European ethnicity on the island prior to Tasman.

In spite of the fact that he had only a partial skull to work with, Dr. Watt's conclusions about the sex and ethnicity of the owner of the skull were later validated by the examination of mitochondrial DNA.

The DNA found in the nucleus of every cell is a combination of DNA from the father and the mother. However, within the cell but outside of the nucleus in a cell are varying numbers of mitochondria, small organelles which perform a variety of important functions, including furnishing energy for the cell by generating the nucleotide adenosine tri-phosphate, or ATP. DNA is found in the mitochondria as well as the nucleus, and unlike the DNA in the nucleus, mitochondrial DNA comes only from the mother (men do not pass on any mitochondrial DNA to their children). Thus, the mitochondrial DNA does not change from one generation to the next, the way the nuclear DNA, mixing the genes of the two parents, changes not only from each generation but even between two siblings of the same generation and parentage.

This fact makes mitochondrial DNA an extremely useful tool. One other important fact makes mitochondrial DNA even more useful, and that is the fact that mitochondrial DNA does change occasionally through mutation. Without this fact, the mitochondrial DNA of every person on the planet would be identical. As it is, there are vast groups of people whose common parentage can be traced by their identical mitochondrial DNA, and other groups whose mitochondrial DNA points to their descent from a mother whose mitochondrial DNA had a small genetic deviation from her mother's at some point in the distant past.

Because of these characteristics, mitochondrial DNA can be used as a very important data point in attempting to identify membership in these broad families, and because scientists can extract and amplify mitochondrial DNA even from bones as old as the skull of the Ruamahanga Woman, mitochondrial DNA is very useful indeed for analysis.

The various groups who share a single common mitochondrial DNA pattern are known as haplogroups (from the Greek word *haplo* or "single," because they share the same gene sequence). At some point far back in the family tree of women in the western Eurasia region, a variation in the mitochondrial DNA of a mother created the haplogroup N, which would then be passed on to all her descendents. Subsequent variations within this broader N haplogroup produced the haplogroup H, which unites a group of descendents of the N-group most common in the region of southern Scandinavia and Jutland.

When cells from the skull of the Ruamahanga Woman were studied, they confirmed the conclusions that forensic anthropologist Robin Watt had made using bone morphology – the skull had belonged to a woman of the H haplogroup. The H haplotype is distinctly European.

Because mitochondrial DNA is only passed from the mother, the presence of the H haplotype rules out any speculation that a male European on one of Tasman's ships had somehow gone ashore and produced a child with a Maori woman, because their child would inherit only the mitochondrial DNA of the Maori mother, which would be the B haplogroup (and specifically the B4a1a1a or "CCGT" haplotype) (Miles Clifford Benson, *Mitochondrial Genome Variation and Metabolic Traits in a Maori Community*).

Of course, judging from the entries in Tasman's journal, such a liaison would be speculative almost beyond possibility, as the European ships were clearly viewed as hostile and armed warriors were stationed along the heights of the western shore to watch their progress and warn the populace (as well as probably to summon a force to repel them) should they have decided to land, which they declined to do. Allowing one of them to even see one of their women, let alone spend some time alone with her, was probably out of the question. In any event, such speculation is cut off by the mitochondrial DNA analysis identifying Ruamahanga Woman as the European daughter of a European mother (her mother could not have been Maori).

Thus, due diligence suggests that the skull is actually that of a

European female, but what of the assertion that this female died during the 1600s?

The dating of the skull to the period between 1619 and 1689 was done at the Rafter Radiocarbon Laboratory, the world's oldest continuously running radiocarbon lab and the first in the southern hemisphere to use carbon-14 analysis. Carbon-14 dating measures the percentage of carbon-14 (also known as radiocarbon, because it is radioactive, decaying with a half-life of 5,730 years).

The vast majority of carbon dioxide in the earth's atmosphere is regular non-radioactive carbon-12. However, some nitrogen-14 atoms present in the atmosphere are converted into carbon-14 by the cosmic radiation striking the upper atmosphere. If an energized neutron (knocked free by incoming cosmic radiation and known as an alpha particle) strikes a nitrogen atom, with seven protons and seven neutrons, a proton will be knocked loose (creating a hydrogen atom) and the former nitrogen atom will capture the neutron, creating a radioactive carbon-14 atom (with six protons and eight neutrons).

Over time, carbon-14 atoms, which are unstable, will undergo beta decay wherein one of the neutrons will split, emitting one electron (a beta particle) and an anti-neutrino, and becoming a proton in the process. This leaves seven neutrons and seven protons, which forms a stable nitrogen-14 atom. Physicists have determined that any quantity of carbon-14 will decay into half of that quantity over a period of 5,730 years, which is the half-life of the radiocarbon.

The earth's atmosphere contains carbon dioxide, which is absorbed by plants during photosynthesis. Some of the carbon they absorb will be carbon-14, in the same proportion as the ratio of radioactive carbon-14 to stable carbon-12 that is found in the overall atmosphere during the time the plants are absorbing carbon dioxide (during the plants' lives). Plant-eating animals will also incorporate carbon into their bodies in the same proportions, through the plants that they eat. Meat-eating animals will incorporate the same ratios into their bodies, from the animals they eat. The ratio is currently about a trillion to one of regular carbon-12 to radioactive carbon-14.

Radiocarbon dating uses this knowledge to extrapolate the age of the remains of plants and animals by measuring the carbon-14 in their remains. Once a plant or animal ceases to live, it ceases to incorporate additional carbon-14 into its tissues through photo-synthesis or digestion. The carbon-14 it absorbed in life slowly decays through the process of beta-decay described above. By estimating what ratio of carbon-14 was present during the life-time of the plant or animal, and by comparing that ratio to what remains today, the rate of decay (or half-life) of the radiocarbon can be used to estimate the amount of time that has passed since the plant or animal ceased absorbing new carbon.

Radiocarbon dating must necessarily make assumptions about the ratios of radiocarbon to regular carbon-12 present in the atmo-sphere in past centuries. There have been significant atmospheric events which have altered the ratios in recent centuries. Most significant of these were the atomic tests conducted during the first decades of the atomic age, during which actual nuclear weapons were detonated in the atmosphere, releasing significantly more radiation than the earth receives from cosmic rays, and thereby producing higher levels of carbon-14 (by a factor of about 100% greater than normal). This "bomb carbon" began with tests in 1955, and peaked in the years 1963-1964 (later in the southern hemisphere than in the northern). All plants and animals alive at that time, as well as all those which have lived since, have absorbed greater ratios of carbon-14 than were present before the atomic testing took place, resulting in a "signature" that marks them as having lived during or after the age of nuclear weapons.

Other changes to the carbon ratios came from the increase in the burning of coal and oil in the Industrial Revolution, during which old coal and oil (in which the radiocarbon had significantly decayed) were burned and their carbon atoms released into the atmosphere, diluting the ratio and lowering the incidence of carbon-14 atoms, thus reducing the amount of carbon-14 absorbed by plants alive after the year 1890 by a measurable amount. Thus, organic material measured after 1890 has less carbon-14 than might be expected, leading to the appearance of greater age (it appears as though more time has passed allowing greater beta decay, whereas in reality the organisms simply absorbed atmospheric levels of carbon that were lower due to the Industrial Revolution).

Certain marine conditions and certain aspects of active volcanic activity also appear to alter the absorption of carbon-14 by plants and animals living in certain niche environments on the globe, leading to unexpected dates in the radiocarbon dating of grasses growing near an active volcano (within 200 meters) or organisms shielded by deep ocean waters or old limestone formations.

All of these examples illustrate the important truth that radiocarbon dating is not a black-and-white, cut-and-dry process that yields an automatic readout, but rather that it – like any other form of analysis – requires the construction of a thesis based upon the available information and certain underlying assumptions. Further, as will be discussed later, there is the possibility that other significant geologic events in the distant past (in particular, the events surrounding the worldwide flood described in similar details by widely dispersed people-groups throughout recorded history) could have altered the carbon-14 ratios in ways that conventional radiocarbon dating assumptions do not incorporate.

However, for organic tissues which ceased to be alive within the past few hundred years, such possible distant events are not a significant factor (since the global flood we will discuss, if it took place, was many thousands of years earlier). Thus, the radiocarbon dating done on the Ruamahanga Woman should produce a fairly reliable date of death. The findings of the Rafter Radiocarbon Laboratory, which uses an extremely accurate method involving an electrostatic Van de Graaff accelerator to separate the atoms of varying molecular weights and which allows the scientists to count the individual carbon-14 isotopes versus the stable carbon-12 and carbon-13 isotopes, discovered that the skull had no bomb carbon signature, meaning its owner ceased to absorb new carbon prior to 1955. Further, the ratios indicated that about 296 years of beta decay had taken place prior to the radiocarbon base date of 1950, plus-or-minus thirty-five years.

Because the science behind radiocarbon dating (at least for dates within the past 5,000 years – more on that subject later) and behind mitochondrial DNA analysis are both so well established, those who wish to contest the conclusion that a woman of European ethnicity was present in New Zealand between 1619 and 1689 do not typically dispute that the skull itself belonged to a European woman who died between those years. That would be a frontal assault on a position that is too well defended. Instead, they have been reduced to completely speculative suggestions with absolutely no historic evidence behind them at all. In fact, the only thing they can offer to support these fantastical theories is their confidence that "Europeans could not possibly have been present in New Zealand" prior to Tasman – their entire framework does not admit to such a possibility.

The leading speculative theories fall into two categories – the first being that an unknown shipwreck must have taken place, during which some female European survivors remained in New Zealand. No record of such a shipwreck currently exists.

The second is that some scientific cadaver or skeleton from the 1600s must have been brought to New Zealand during subsequent centuries and then lost, the skull of which traced an unknown odyssey that ended along the banks of the Ruamahanga in the year 2004. Again, no record of such a missing cadaver exists. Further, the region through which the Ruamahanga River flows is very rural, and sparsely populated, and it is difficult to concoct a scenario by which a scientific cadaver would be lost upstream from the discovery point. The Ruamahanga flows down from some very rugged and wild mountains before entering the broad Wairarapa Valley, but none of these areas are home to a museum or university. Further south and to the west, the city of Wellington might be a candidate for a cadaver for study, but such a cadaver would very likely have arrived on a ship from the south, and if lost the skull would certainly not have rolled to the north, over the steep Rimutaka Ranges, and into the Wairarapa. Thus, even these speculative theories (with no historical evidence to back them up whatsoever) come apart under closer scrutiny.

If there were no other archaeological and anthropological evidence that "Old World" people-groups had been to New Zealand before the Maori arrived sometime in the thirteenth century (roughly between AD 1200 and 1300), then such flights of speculative fancy might possibly be justified to explain away a single outlier, even an outlier that is as full of indisputable evidence as the Ruamahanga skull. However, there are many other corroborating data points that can be piled up, and that we will examine in this book. There are so many that we will not be able to examine them all, but we will examine enough to exhaust the likelihood that our thesis is mistaken. In fact, the Ruamahanga Woman is just one of countless pieces of evidence – very strong evidence – from around the world that the current conventional framework of anthropology, biology, and geology is wrong.

The Celestial Phenomena and the

Clues to Mankind's Past

If Ruamahanga Woman is not the product of some postulated "lost medical skull" from the 1800s (and there is no evidence to suggest that she is – such a story is a complete fabrication invented to prevent the unraveling of the dominant biological, anthropological, and geological models), then what are the alternatives?

The alternative which the guardians of the conventional explanations wish to avoid is the possibility that ancient people groups currently excluded from history were able to (and regularly did) traverse the oceans, a feat requiring technology and navigational skill (including perhaps geodetic knowledge of the size and shape of the earth) in ancient cultures that goes far beyond anything that mainstream academia will countenance.

Indeed, the conclusions to which Ruamahanga Woman could lead pose a grave threat to the neat linear progression favored by anthropologists who (in line with Darwinian assumptions) postulate steady advance from primitive hunter-gatherer societies towards the more advanced cultures of Sumer, Babylon and Egypt, and then ancient Greece and onward into recorded history. The ability to navigate the globe as far as the remote south Pacific by "Old World" peoples would surpass any known technology or science achieved by any of the ancient civilizations, at least in the form in which they are currently described in the conventional texts.

Such a possibility would also threaten numerous politically correct sacred cows, particularly if it were demonstrated that cultures from the "Old World" influenced cultural practices and behaviors in the New, including those in Central and South America and as far away as New Zealand and Australia. It is automatically (though wrongly) assumed that such a possibility would somehow "take away" from the achievements and culture of the Maya, the Aztecs, or the Maori – an unspeakable heresy in conventional circles and a possibility many interest groups will resist with every fiber of their being.

But a single skull found along a river in the Wairarapa (even one so full of incontrovertible evidence as the skull of the Ruamahanga Woman) is easy to discount as an outlier, an accident of history, the product of a preposterous story about a medical specimen, or some other unsupported explanation. If there is only one outlying data point and an overwhelming amount of data points for the other side, then even a preposterous story might be the real explanation for the Ruamahanga skull. The question then becomes, "Are there other data points which support the heretical conclusions hinted at by the data contained in the Ruamahanga skull?"

There are other data points – a vast array of them, in fact. There are two extensive reservoirs of clues from the ancient world, and both of them appear to point to exactly the level of global dispersion and precise geodetic and astronomical knowledge described above, knowledge beyond anything the conventional models ascribe to even the highest achievements of ancient Greece or Egypt. These two extensive reservoirs of evidence remaining from mankind's distant past are the myths found around the world, among the descendants of civilizations separated by vast oceans, and the architectural structures and other forms of archaeological evidence found throughout the world (including not just structures but smaller artifacts, inscriptions, and even human remains), again in locations separated by vast oceans.

To fully appreciate the weight of the evidence, it is necessary to understand the phenomena which figure prominently in both worldwide myth and worldwide ancient architecture. Those phenomena are the celestial mechanics of the stars, the planets, the annual path of the earth, the tilt of the earth's axis, and the ages-long motion of the earth's axis and poles known as "precession."

These celestial mechanics are treated in various texts, some of which include some good diagrams, but almost all of which stop short of spending the time and space necessary to make all of the moving pieces completely clear.

The best diagrams are perhaps those created by H.A. Rey (1898 – 1977), the beloved children's author and creator (along with his wife Margret Rey) of the Curious George stories. In his 1966 text *The Stars: A New Way to See Them,* he employs his lucid prose and ample artistic ability to explain these phenomena in adequate detail for the reader to come away with a clear picture of precession and the impact of earth's orbit and axial tilt upon the movement of the constellations through the sky. However, because he was explaining the heavenly mechanics primarily for the purpose of stargazing and not for the purpose of examining the

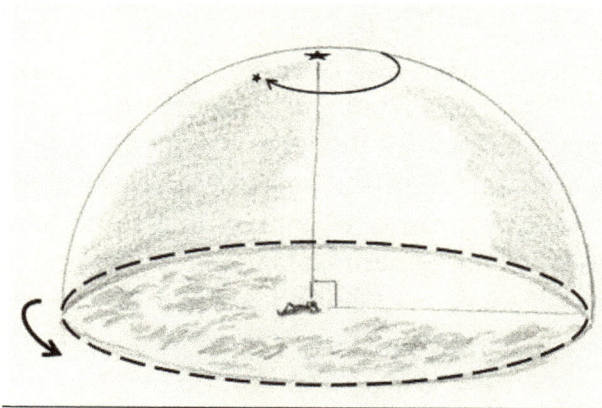

To an observer lying at the earth's north pole, looking directly upward, the entire sky will turn about a central point directly overhead, just like the central point on an old record player or the axle of a wheel. This central axis point of the sky will not appear to rotate, and a star located at that point, which is called the celestial north pole, will not move during the course of the earth's rotation. Polaris, the north star, is currently at this location in the sky (or rather, extremely close to it, and close enough to serve as a marker of the celestial north pole). The direction that the sky will appear to turn will be the opposite direction that the earth itself is actually spinning (just as objects outside a moving car appear to be moving in the opposite direction that the car is moving).

vast repository of ancient myth and ancient architecture, even his outstanding diagrams do not always go far enough.

In this chapter, I will present diagrams, some of which are conceptually descended from those drawn first by H.A. Rey, which should help lay the foundation for the reader to understand the phenomena which myths from around the world, and ancient architectural remains on several continents and Pacific Islands, encode and preserve to this day.

Modern man has become generally unfamiliar with these phenomena (which may, in fact, never have been widely known to the "general populace" but guarded carefully as special knowledge for an elite few), and because of this general unfamiliarity, the amazing body of evidence that supports the wildest implications of the Ruamahanga skull goes largely unnoticed and unexamined.

This is similar to the situation that would ensue if one side in an argument about some important subject of current events refused to use computers or the internet or mobile devices of any sort, and thus were completely unaware of the flood of voice and video data surging all around them, carrying conversations and data and evidence that they should know about, but which would remain completely invisible to them because they refused to make use of the sensors which would allow them to hear all those voices.

If we desire to "tune in" to the voices all around us in the myths inherited from ancient millennia (many of them quite familiar, such as the Greek myths and the Norse myths), and to "tune in" to the voices in the structures left behind by the ancients (again, many of them familiar as well, such as the remains at Stonehenge or the pyramids of Mexico), we must gain a working knowledge of the celestial mechanics which were clearly understood by the ancients and which clearly were held to be of tremendous importance.

To begin, let us imagine that we are standing outside, observing the stars overhead. The best place to begin might be at the north pole itself, where the motion of the earth and the motion of the sky are perhaps easiest to conceptualize.

If one night you were to lie facing up, right at the north pole, and observe the stars overhead, you could observe the apparent "turning" of the entire vault of the heavens, due to the rotation of the earth.

Because the point around which all of the earth's turning takes place would be right below your head as you rested on the ground – or ice sheet, in the case of the north pole (see diagram above) – you could easily imagine that you were on a vast disc which was turning like a record on a record player (for those who remember such devices), or a CD inside a CD player, or even the lower stone in a great stone mill-wheel (for those unfamiliar with the mechanics of a typical water-driven stone mill from the days before electricity, a diagram of such a machine appears below – it is introduced here because, not having record players or compact discs that we know of, the ancients used this metaphor exten- sively as a mythological reference to the great machinery of the sky). In the above analogies, your point of view would be from the center post around which the record spins, or from the central hole found in the middle of your CD, or from the hub in the center of the stationary lower millstone, looking up at the eye in the upper "runner" millstone .

Because of our perspective as tiny beings upon the enormous globe on which we are riding, we do not tend to feel the earth's motion or perceive it as turning (in fact, it is so large we do not even perceive it as a globe). Instead, we perceive the motion of the background that appears to be moving instead – exactly the same way that it is possible to be sitting in a big passenger jet on a runway or at the terminal gate for a long period of time, which begins moving without any announcement, and to suddenly feel that the buildings or landscape you see out the window are moving, while you do not initially realize that it is the airplane that has quietly started to move (the same thing can happen in a train). Thus, if we were reclining at the north pole, although the world appears to be a circular disc, we don't perceive it to be spin- ning; instead, we perceive the sky to be spinning, in exactly the same way that we can perceive the buildings to be moving past us in a plane or a train when in fact it is they which are standing still and we are the ones moving instead.

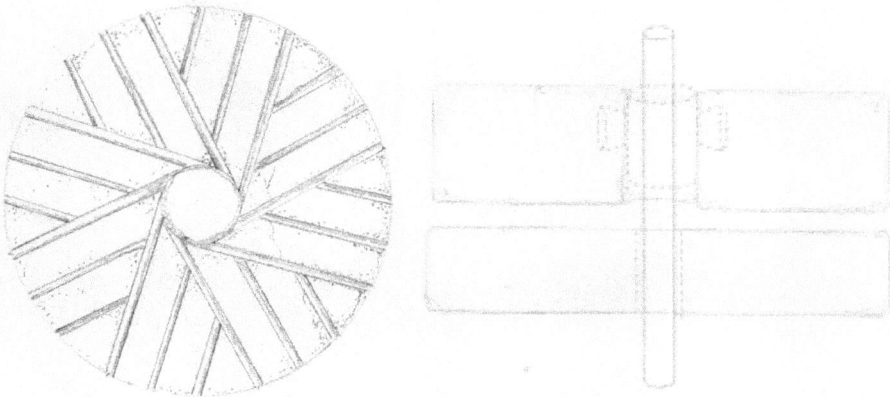

Illustration of a typical millstone and configuration. The left illustration shows the cutting surface with furrows, which were cut into both the lower face of the upper "runner" stone and the upper face of the lower "bedstone." The right side-view cutaway illustration shows the lower bedstone, which does not move, and the upper runner, which is suspended from the central spindle by an "iron cross" shaped device known as the ryrd. The turning spindle will thus turn the runner but not the bedstone. Note the aptness of the millstone metaphor for the turning heavens, and the similarity to the previous diagram of the observer lying at the north pole looking up.

Thus, in the first diagram showing the observer at the pole, we have depicted the earth as it appears to us on the ground (as a flat disc) and the heavens as they appear to us from the earth – as a kind of "snow globe" over our heads, the inside of a huge sphere, of which we can see only one half (or hemisphere) at any time. As depicted in the diagram, this celestial hemisphere is what actually appears to be turning, from the point of view of our observer on the ground. And, just as the buildings we see out the window of our train or taxiing airplane appear to be moving in the opposite direction from our true motion (moving backwards when the train or plane is moving forward), the inner part of that imaginary hemisphere or snow globe will appear to be rotating in the opposite direction from the actual spinning of the earth that we are "riding on."

Of course, there are no real buildings in the sky for us to perceive its turning motion, and so the only way we really perceive it to be

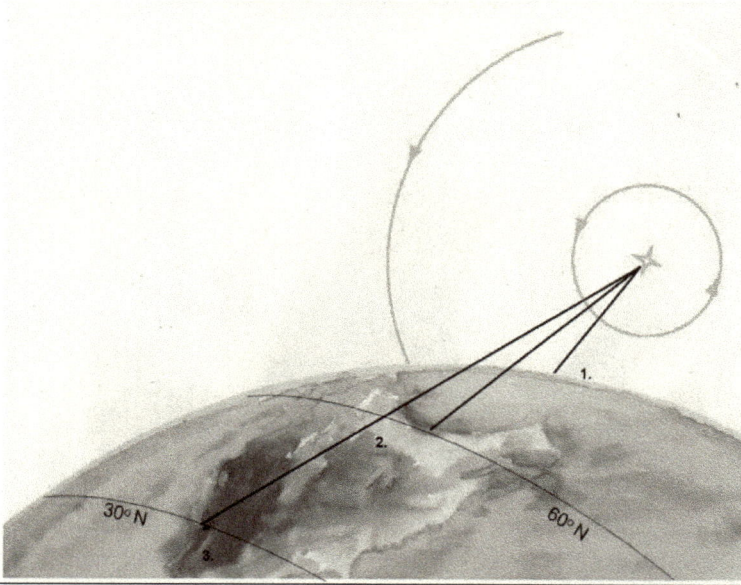

The angle of elevation of the pole star corresponds to the latitude of the observer. At earth's north pole (position 1 above), which is latitude 90° north of the equator, the pole star is at elevation 90° (straight overhead). As one moves south to latitude 60° and 30° (positions 2 and 3 above) the pole star will sink 1° for every degree they proceed south, so that at 30° north latitude the star will be at elevation 30°.

turning is if we observe the stars, which are only visible during the night, or the sun and the moon. The apparent motion of the sun, moon, and stars each day is primarily due to the rotation of the earth. From the perspective of the observer at the north pole, the stars will appear to turn in perfect circles around a central

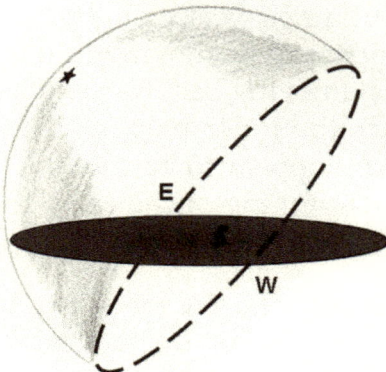

As the observer moves south from the north pole, the north star sinks in the sky behind him, and the celestial equator tilts up from the horizon to arc across the sky, 90° "down" from the celestial north pole.

hub directly overhead, just as little dots of paint would appear to turn in circles if we placed little white paint dots on a record and then put it on a record player (of course, the record would be upside down in this case, which is another reason the analogy of the upper and lower millstones is so powerful – the heavens can be thought of as the underside of the upper, rotating runner stone, seen from our vantage point upon the lower, stationary bedstone).

There is one star which will not follow a perfect circle around the pole star, even to an observer located at the north pole, and that star is the sun.

If the earth turned on an axis which was perfectly perpendicular to the plane of its orbit around the sun, then to an observer at the north pole, the sun would appear to roll along the dotted line of the horizon in the initial diagram every day of the year.

However, the axis of the earth is inclined at an angle of 23.5° from the plane of the earth's orbit around the sun (more precisely, it is inclined at an angle of 23° 27' but for the present discussion we will use twenty-three and a half degrees for simplicity). Therefore, the sun at the north pole stays above that dotted blue line all day long when the north pole is inclined towards the sun, and stays below it when the north pole is pointed away from the sun. The impact of this axial tilt will be discussed in greater detail shortly.

Since next to nobody actually lives at the north pole itself (or the south pole either), we will now proceed southward from the pole to observe the effect this will have on the position of the celestial north pole and the celestial equator. As we travel southward, we can see that the celestial north pole will no longer be directly overhead, but will sink in the sky by one degree of elevation for each degree of latitude we move southwards. The lower diagram at left helps conceptualize why this is so.

Similarly, as the celestial pole sinks lower, the celestial equator which remains 90° "down" the imaginary celestial hemisphere from that celestial north pole will appear to "rise" up from the southern horizon even as the north celestial pole sinks towards the northern horizon. The diagrams at left depicts what happens to the celestial "snow globe" or hemisphere as we proceed south from the north pole – it "tilts."

It is now possible to discuss in greater detail the path through the sky traced by the sun as the earth rotates about its axis each day.

The diagram at right below shows the big picture of the impact of the axial tilt as the earth proceeds around the sun. Notice that as the earth proceeds around the sun, the axial tilt of the earth remains constant (ignoring for the moment the eons-long motion of the axis itself, known as precession and discussed in greater detail later). Thus it remains pointed at the pole star, Polaris, which is marked in the diagram. Because Polaris is actually so much farther away than its depicted point on the vault of the sky in this diagram, the axis points to it at all times, although in the diagram the axis of the earth as depicted at various stations on its orbit may not.

The diagram also indicates the celestial equator which we saw in the previous "snow globe" and "tilted snow globe" diagram. You can think of the celestial equator in the diagram below as forming the joint where the two celestial hemispheres come together (the glass of two snow globes so to speak). The celestial equator is the circle where the upper (northern) celestial hemisphere and the lower (southern) celestial hemisphere fit together to form a full sky globe.

In the above diagram, it is possible to observe that the path of the sun will appear to a viewer on the earth to follow the dotted line designated as the "ecliptic," which corresponds to the plane of the earth's orbit (and the plane of the orbit of the other planets) around the sun. For instance, looking at the earth's location on December 23, an observer on the earth looking towards the sun will create a line of sight that goes from the earth to the sun in the diagram at right, which could be extended to the "vault of the sky" represented by the globe (the two halves of the celestial hemisphere described above as the glass in an imaginary "snow globe") and this line of sight would trace out the dotted line marked as the ecliptic.

It is readily apparent that the ecliptic will be "above" the celestial equator for half of the year, and "below" it for the other half. More precisely, it will be above during the day and below during the night for one half of the year, and then below during the day and above during the night for the other half. To help conceptualize what this means for an observer on the earth, first consider the view from the earth at the June solstice. When the observer is on the side of earth turned away from the sun (night at his location) he will see celestial objects on the ecliptic (the zodiac constellations and the visible planets) below the celestial equator. When earth turns and the observer faces the sun (day at his location) he will see the ecliptic path of the sun trace an arc above the celestial equator.

Remember that the celestial equator is always 90° from the celestial north pole: it does not change during the year for an observer at the same latitude. It is the tilt of the earth which causes the sun's path (the ecliptic) to be north of the celestial equator in the summer and south of the celestial equator in the winter. The ecliptic will cross to the north or the south of the celestial equator on either equinox, and spend half the year moving north of it and half moving south of it.

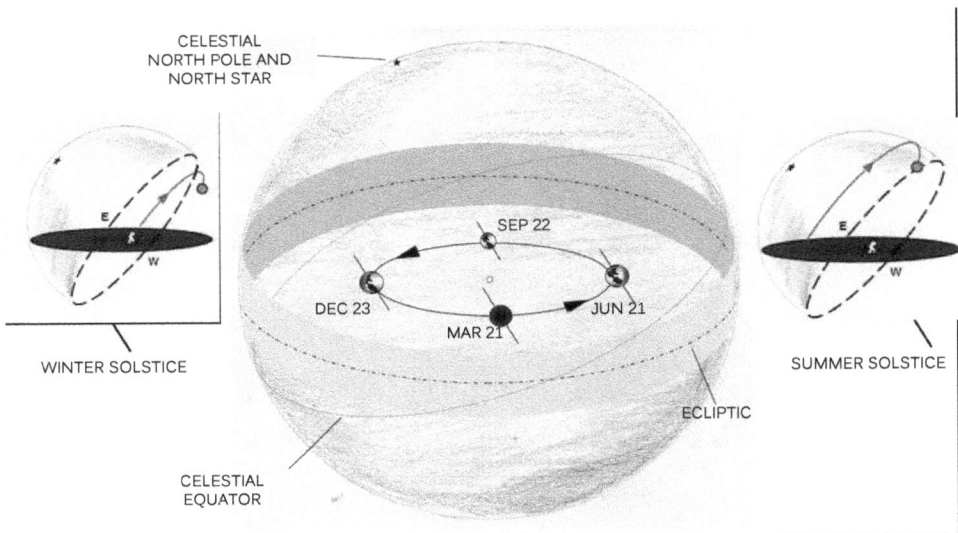

From the earth's position at winter solstice, the ecliptic path is below the celestial equator. This can be seen in the central diagram of the earth's orbit, if you imagine an observer looking across through the sun and seeing the ecliptic below the celestial equator. In the "tilted snow globe" the sun's path through the sky (ecliptic) is also below the celestial equator. At summer solstice, the opposite is true.

The same annual movement of the ecliptic path between the solstices and across the celestial equator at the equinoxes is depicted once again below, but this time from the perspective as someone actually standing on the ground watching it. This presents the same phenomenon of the shifting ecliptic, but from the more familiar vantage point that we are accustomed to seeing each day.

This illustration shows the same phenomenon depicted in the previous diagram, but from the observer's point-of-view, "in 3-D." The vantage point is of an observer on the ground facing north. In order to depict the angle of the ecliptic path along which the sun travels, this drawing imagines the observer inside a "shadowbox" or a room with four walls. Remember that the path of the sun on the two equinoxes will be along the celestial equator. On the summer solstice, the path will be above or north of the celestial equator, and below on the winter solstice. Arrow depicts earth's rotation towards east.

The importance of this steady movement of the sun's rising from one solstice to another in a regular, reliable, annual round trip is emphasized in countless existing archaeological sites around the world, from Stonehenge to Tiahuanaco. The crossing of the ecliptic and the celestial equator are likewise depicted on statues and art worldwide, as we will discuss later.

As important as this annual movement is, however, and as important as it is to be able to understand the cosmological reasons which cause it to take place, it is only the tip of the iceberg. Behind the rising of the sun, and blotted out each day by the overpowering light of that closest star, are countless other stars. These will also appear to move due to the daily rotation of the earth and due to the earth's progress around the sun. The stars which blanket the vault of heaven thus provide further "handles" for observers on the earth to use in order to grasp the motion of the celestial machinery and to perceive annual patterns even more sophisticated than the rising points of the sun.

We have already made some reference to some of the stars which help us observe the effects of the rotation of the earth, in particular the North Star which currently marks the celestial north pole and thus the point in the sky around which the rest of the entire vault of heaven appears to rotate.

As we have already seen in the first diagram depicting an observer at the north pole, stars further away from the celestial north pole will turn in a circle around it, the circle being larger the farther away the star is from the central hub of the sky's apparent rotation. As we saw, this sky rotation is an illusion, caused by our own rotation on the earth, just as stationary buildings will appear to move (in the opposite direction from our direction of travel) when we look out the window of a slowly taxiing jumbo jet or a moving passenger train.

The further we move down the globe, the lower in the sky the celestial pole will appear, and some of the larger circles made by stars situated further out from the central point will be intersected by the horizon each day, obscuring that star from our view until the earth turns around enough to enable us to see it again.

This obscuration of part of the "circle made by the star," in fact, happens every day for that nearby star known as the sun. Of course, because of the tilt of the earth's axis, if you were within 23.5° of the north or south pole, then the sun itself would fall into the category of stars which do not appear to set but rather perform a full circle in the sky, and then the sun would not set at all during the summer portion of the year. Everywhere else, the sun's "circle" dips below the horizon for a portion of the daily turning of the earth.

The diagram below illustrates, from the perspective of an observer on the ground and in the northern hemisphere, this stellar motion. Stars far enough away from the pole star rise in the east (as the earth's curving surface speeds in that direction, with all the human observers on it) and progress overhead to set in the west, but stars closer to the pole star make a smaller circle and do not actually rise or set, but turn in a counter-clockwise direction. In fact, the stars (and constellations) that do rise and set can be thought of as turning in their own counter-clockwise circles around the pole star as well, except that their circles are much larger – so much

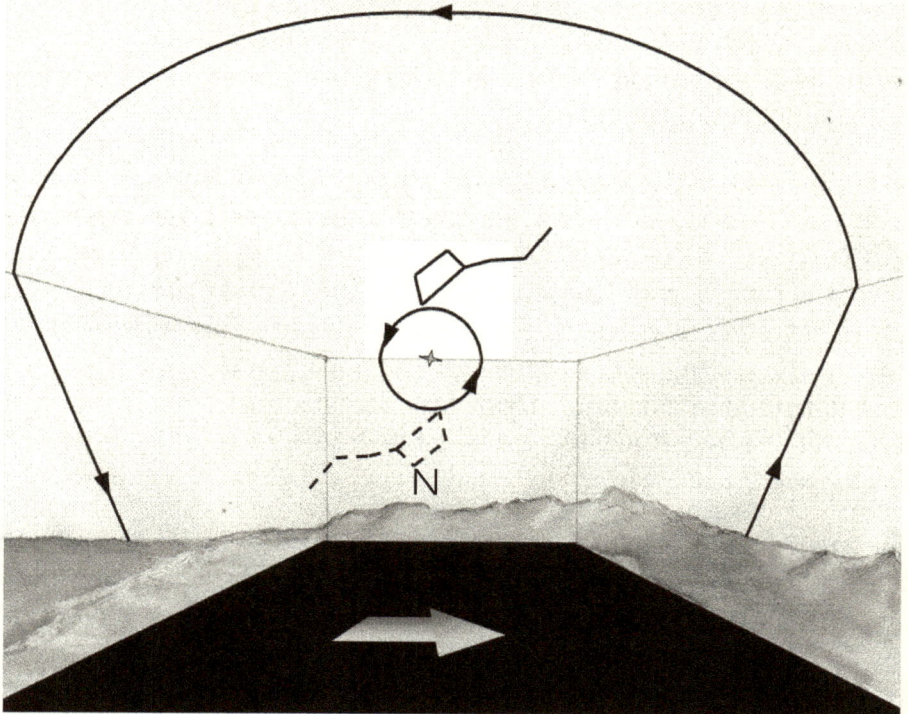

larger that they are blocked from sight by the rest of the earth for part of their journey, and thus appear to "rise" and to "set."

The sun's motion against terrestrial landmarks never changes, but will rise in the same location on the same day each year (if the observer is able to ensure that his vantage point is exactly the same as it was on previous years). Its only motion is its endless back-and-forth circuit between two identical endpoints. However, when the background of the stars is added to the picture, a referent beyond the earth and the sun comes into play, adding an important new dimension to the picture.

In fact, the set of stars that appear to be "behind" the sun (or in the vicinity of the sector of the sky from which the sun is rising) will necessarily be different as the earth progresses in its annual orbit around the sun (see diagram below).

Thus, if you look east from your vantage point on the curve of the earth that is turning along the earth's axis towards the east, each night the constellations along the ecliptic rise from the east and make their way across the vault of the sky, sinking into the west, where they continue their counterclockwise circle out of sight.

An excellent way to conceptualize the celestial mechanics being described in the diagram on the following page is to imagine it taking place in a familiar room in your own home. If you think about the earth's orbit around the sun, and imagine it takes place in your dining room (for example), the stars in the center of the ceiling would be visible all year around for an observer in the northern hemisphere, but the stars on the walls would only be visible on certain parts of the earth's circuit.

Because we only see stars at night, an observer on an earth orbiting inside your dining room would only see the stars on the walls they were passing, when that observer was turned away from the sun in the center of the room. The stars on the wall across from the sun would be obscured by the sun until the earth made its way around to the other side of the room, at which time those stars would be visible to observers on the side of the earth that was turned away from the sun (which happens every night as the earth spins).

Because the earth is progressing by about 1° on its complete circle each day, it is in effect "passing" the stars on the walls as it moves along. The net effect of this forward progress is that the stars rise four minutes earlier than they did the previous night. For example, as the earth makes its way around the sky, Orion rises and sets four minutes earlier each day, until he is rising during the day. His setting will also be earlier and earlier, and as spring moves towards summer, he will begin setting closer and closer to the time that stars become visible. By late May, he will only be visible in the early evening, when he is getting ready to set in the west.

As the earth continues around the sun and his rising points get earlier and earlier, he will rise and set completely during the day, until his rising becomes early enough to be seen low in the sky prior to the sunrise: an important date of return and a phenomenon known as heliacal rising (a term derived from the name of the ancient sun god Helios).

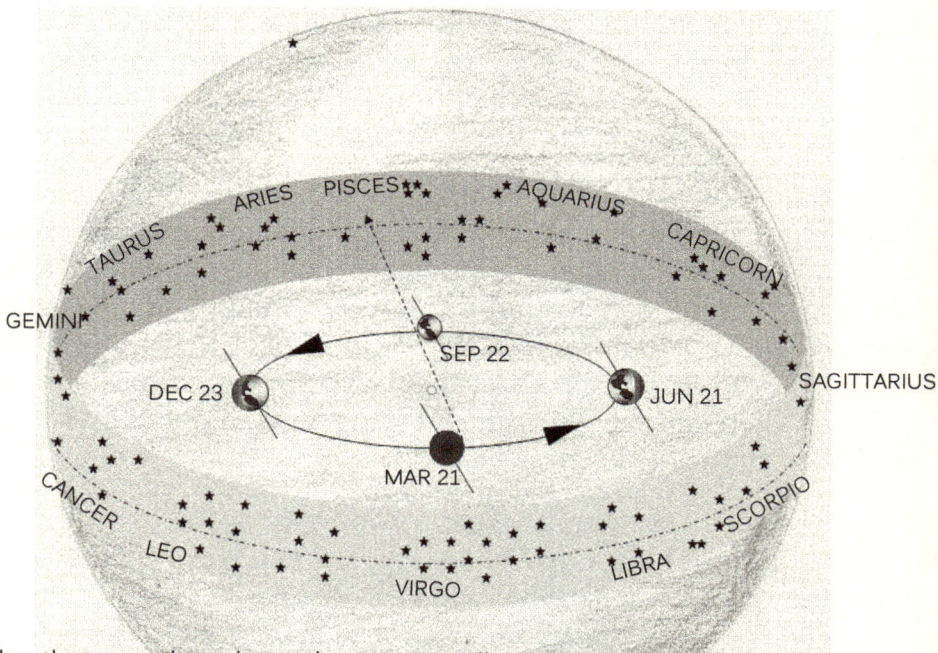

Why the sun rises in various constellations throughout the year: as earth turns on its axis, it reveals different constellations during the night. The last ones its turning reveals before the sun is revealed differ as the earth makes its orbit around the sun. At March 21 the last stars revealed before the sun "comes up" will be those of Pisces (as indicated by the arrow from earth "across" the sun to the other side).

As the eastern edge of the earth turns around to reveal the first rays of the sun, the group of stars in the sky just before the sun rises (stars in their heliacal rising) will reveal that the earth is moving in its orbit around the sun, and where it is in that orbit.

The constellations or groupings of stars are not actual groupings in outer space but appear to be grouped together from the vantage point of earth as if they were arranged on a two-dimensional background, even though in reality some are far closer to us than others that appear to be their "neighbors." Those groupings which are aligned with the plane of the earth's orbit, and therefore coincide with the path that the sun takes through the sky (the ecliptic), are called the constellations of the zodiac. They are drawn in the diagrams as if they were in a belt running outside the orbit of the earth.

Each morning, as the sun appears from a different point on the eastern horizon, it rises out of a section of this zodiac belt with a particular constellation that can be predicted based upon where the earth is on its orbit. If you were to draw a line from the earth to the sun (or, more precisely, from that point on the leading edge of the earth as it spins and therefore where the sun appears to rise for those standing on the spinning ball of the earth) and then continued past the sun to the constellation in the zodiac belt that is opposite the earth, you would see which constellation was in the eastern part of the night sky just before the sun appeared above the eastern horizon each morning.

Up to this point, the celestial phenomena which we have been discussing have been phenomena which can be observed with the naked eye by anyone who chooses to take the time to carefully observe the heavens over the course of a single year. The constellation dominating the section of the sky from which the sun rises on each of the four important points of the earth's orbit described above – spring equinox, summer solstice, autumnal equinox, and winter solstice – are the same each year, and would be forever, but for a truly remarkable celestial phenomenon, a phenomenon which is almost imperceptible, and which in fact moves so slowly that a single human lifetime – even an exceptionally long and healthy human lifetime – is arguably too short to perceive it at all.

Currently, the sun will appear out of the constellation Pisces on the morning of the spring equinox each year. The spring equinox in many ancient cultures around the globe marked the beginning of a new yearly cycle (the month of Passover is calculated each year based upon the date of the spring equinox, for example, albeit via a rather complicated formula which takes into account many other factors, with the date of the spring equinox one of the most important of these many factors). Thus, the constellation from which the sun appears to rise on the morning of the spring equinox is of great importance to our discussion.

The other constellations which will see the sunrise on the summer solstice, autumnal equinox, and winter solstice if the sun is rising from Pisces on the spring equinox will be the constellations of Gemini on the summer solstice, Virgo on the autumnal equinox, and Sagittarius on the winter solstice, followed by Pisces on the spring equinox again.

All of this has the comforting predictability of a great precision mechanism, an intricate and accurate Swiss watch in the heavens. The sun rising marches north to the summer solstice and then back south to the winter solstice each year, passing through the same point on the eastern horizon on the spring and fall equinoxes with perfect regularity. The constellations wheel overhead each night along their appointed paths, with the constellations of the zodiac turning along the band through which the sun traces its path during the day. The wheeling constellations are different during different times of the year (because the earth is at a different point on its annual orbit and thus the stars in the "backdrop" of the heavens will be different as one looks from the earth each evening), but there is perfect predictability as to which constellation will occupy the sky where the sun will rise each morning, in an annual procession which (it would seem) never changes.

On the critical days of the spring equinox, summer solstice, autumnal equinox, winter solstice, and then spring equinox again, the predicted constellations of the Pisces, Gemini, Virgo, Sagittarius, and back to Pisces will always be there, like guardians of an ancient and unchangeable order -- except that this order is slowly shifting, as we will discuss in a moment.

In many ways, the face of an analog watch or clock is very much like a model of the turning earth, seen from the vantage point of an observer directly above the north pole. If you stare down at the face of any watch or clock, you can imagine you are staring down at the top of a globe in which the center of the dial is the north pole. Now, if you can imagine that the hands of the watch are standing still, and that the face of the watch begins to move instead, you will see that the face of the watch must move in the same direction that the earth turns, in order to keep the "apparent motion" of the hands going in a clockwise direction. In other words, a watch face is like a model of the turning earth, but instead of creating a face that rotates in a counterclockwise direction, it has "markers" along that face which turn in a clockwise direction, simulating a face which turns in the same direction that the earth itself turns (in other words, towards the east – if you have trouble thinking of this, imagine North America and the other continents of the northern hemisphere depicted on the watch dial; if you are a US

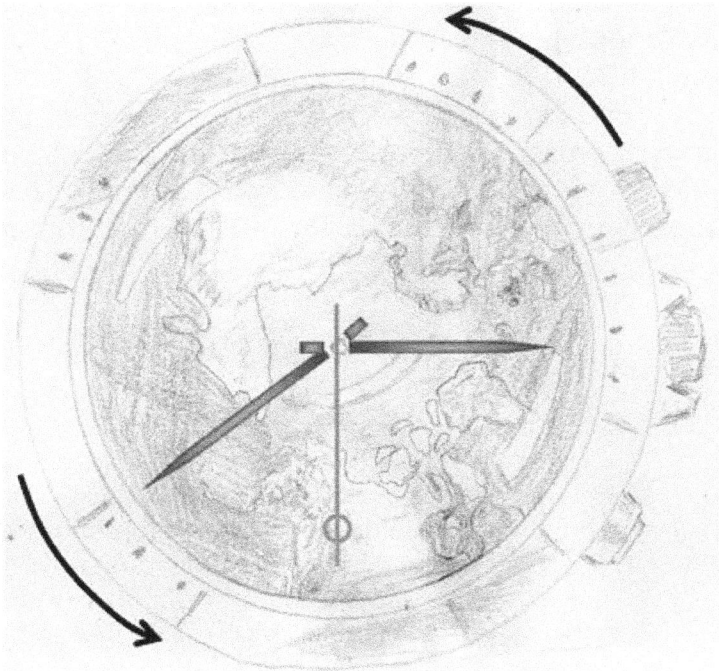

How modern watches are models of the earth's turning: imagine the earth is turning in the direction of the arrows, and the hands are standing still. This illustration clearly shows how our concept of time is derived from the celestial motions.

resident, you can easily remember that the sun rises first over the east coast where New York and Boston are located, and later over the western states, where the time zone is generally three hours later).

The point of this exercise is to realize that, for devoted observers of the heavens (as mankind has clearly been for millennia), the precision of the mechanism of the circling skies is every bit as comforting and reliable as the finest watch – except for one catch. In fact, it is a very big catch, and the catch is that the precision of the heavenly mechanism has apparently been knocked slightly off kilter, such that everything is reeling slowly, very slowly, off course and the constellations that guard the four orbital points are changing almost imperceptibly but surely, and will be replaced by new members of the zodiac every 2,160 years or so.

The cause of this phenomenon is the fact that the earth spins around an axis which is not only tilted 23.5° from perpendicular (perpendicular in relationship to the disc described by its orbit around the sun) but which itself *precesses* or draws a circle of its own, while maintaining its 23.5° tilt (actually, even this tilt "nods" ever-so-slightly, a motion which is known as *nutation*, but which has far less impact on the observations of the stars and the star background of the rising sun that we have been describing above than does precession, and which therefore will not be discussed further here).

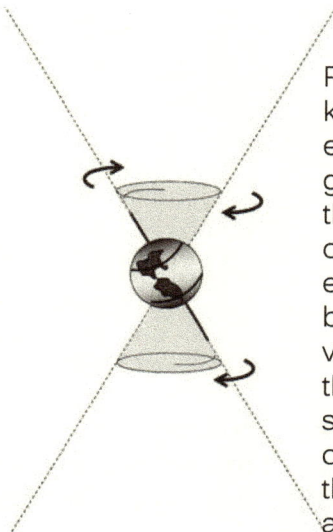

Precession of earth's axis: If earth's axis kept the same angle at all times, then as earth made its annual orbit, the same background of stars would always be seen from the same place on earth at the same time of night on the same day of the year. However, like a gyroscope or spinning top that begins to wobble, earth's axis traces out a very slow circle -- actually two circles, since the axis of the south pole traces one at the same time. The rate of this axial wobble is only 1° every 71.6 years, and it means that the constellations seen on a particular date and time shift 1° every 71.6 years as well.

This axial motion is often described as being like the motion of a spinning top once it begins to wobble (the top at first spinning on its axis and the axis remaining almost perfectly vertical, but then later the axis itself begins to wobble in a circle, while the top continues to spin around this wobbling axis), or like the motion of a gyroscope with an axis which is tracing out a graceful circle even as the gyroscope itself rotates on that axis.

H.A. Rey described the phenomenon this way in *The Stars: A New Way to See Them*:

> The celestial pole is the point in the sky to which the axis of our rotating planet points and the center around which the sky appears to rotate. If the earth were a perfect sphere, this axis would always point to the same spot. However, the earth is not a perfect sphere but is a little flattened at the poles and a little thicker at the equator, and this causes the axis to wobble the way a slowing-down top does. While the tilt of the axis to the earth's orbit remains the same (deviating 23 ½° from vertical), the axis itself describes a funnel-shaped motion, once around in about 25,800 years. If we made a model of this setup, and the earth's axis had a pencil point, that pencil point would describe a circle on the vault of our model sky. 128.

The direction that the axis travels in its slow, once-in-almost-26,000-years circle is opposite to the direction that the earth itself is spinning on its axis. Thus, as the axis proceeds along its very slow circle, the background constellations which sit behind the rising sun on the important orbital stations described above will shift as well, but in the opposite direction that they move through the sky each night (since their nightly appearance of movement is caused by the daily rotation of the earth and that rotation is opposite to the direction of precession). The fact that this motion brings the ecliptic into the *preceding* zodiac constellation (rather than into the successive zodiac constellation as would be expected from the nightly march of the zodiac constellations) gives rise to the term "precession" – it actually describes the movement of the equinox rising point to the preceding constellation in the zodiac.

The impact of this slow circle of the earth's axis is that the constellation behind the sun as it rises on the morning of the spring equinox, which currently is Pisces, is not in the exact same spot each year, shifting about one degree in the sky towards the west every 71.6 years (one degree of a three-hundred-sixty degree circle). It also changes the point in the sky around which the sky appears to turn (the celestial north pole, for observers in the northern hemisphere, and the celestial south pole, for observers in the southern hemisphere).

Thus, over a very long period of one-degree creeping, the entire section of the sky inhabited by the constellation Pisces will be delayed from its expected position behind the rising of the sun on the vernal equinox, and the constellation in the zodiac that rises *before* the delayed constellation (the preceding constellation) will be there instead. This constellation is Aquarius. This phenomenon explains the concept of the arrival of a new age, in this case the Age of Aquarius, each "age" being dominated by four different constellations and named for the constellation behind the sunrise on the vernal equinox.

Obviously, when a new constellation is behind the sun's rising point on the vernal equinox, there will be a new constellation behind its rising on the other important orbital stations of summer solstice, autumnal equinox and winter solstice as well, since the entire belt of the zodiac is basically fixed like the watch bezel to which we compared it previously. The current regime of Pisces (vernal equinox), Gemini (summer solstice), Virgo (autumnal equinox) and Sagittarius (winter solstice) will be replaced by Aquarius (vernal equinox), Scorpio (summer solstice), Leo (autumnal equinox) and Taurus (winter solstice).

Some argue that this has already taken place – room for argument is provided by the differences in the way one divides up the sections of the sky belonging to each constellation and where exactly you draw the boundary between one constellation and the next in the belt of the zodiac. There is evidence that ancient civilizations calculated each of these ages to last 2,160 years, which leads to an estimate for the entire precession of the equinoxes of 25,920 years – a number which rounds the actual 71.6 years required for a degree of precession to 72 years, a very close approximation. When multiplied by the three hundred sixty degrees of a circle, an approximation of 72 years per degree yields 25,920 years. Allotting 30 degrees of arc to each of the twelve constellations (three hundred sixty divided by twelve) would yield 2,160 years to an age.

In the age before the current regime, the four orbital stations were dominated by Aries (vernal equinox), Cancer (summer solstice), Libra (autumnal equinox), and Capricorn (winter solstice). As H.A. Rey and others have pointed out, the description of the tropic lines (Cancer and Capricorn) are relics of this previous age, named for the constellations which dominated the solstices in the previous age. The tropic circles are situated 23.5 degrees north and south of the equator, and define the most northern and southern latitudes at which the sun's rays beat down vertically, doing so at the most northern latitude only on the summer solstice in the north (dominated by Cancer in the previous age) and the winter solstice in the south (dominated then by Capricorn).

If these gate-keepers changed over to the current regime (which is coming to an end or already ended in the opinion of some) about 2,160 years ago (somewhere around the time of Christ or 150 years prior), then that age itself had been in place for another 2,160 years or so, meaning that the previous shift also took place in accessible history – distant, for certain, but not completely unknown – somewhere around the year 2300 BC, give or take some period of time on either side.

The constellations which were replaced by the series led by Aries would have been Taurus (spring equinox), Leo (summer solstice), Scorpio (fall equinox), and Aquarius (winter solstice).

The next series of four diagrams illustrates this important concept, using the metaphor of a great watch bezel, which turns to indicate a new age. The turning watch bezel represents the turning of the celestial north pole and celestial equinox, which is caused by that double-cone-shaped wobble of the earth's axis.

Since, as we have seen in previous diagrams, the locations of the celestial poles and celestial equator are caused by the attitude of the earth's own poles and equator (being projections in the sky of the motion of the earth around its poles), the precessional wobble of the earth will move both the celestial north pole and celestial equator. The starry backdrop, in which the constellations repose, will not move. Hence, in these diagrams, the bezel rotates but the constellations themselves do not.

Along the watch bezel are marked the locations of the vernal equinox (VE), the summer solstice (SS), the autumnal equinox

AGE OF TAURUS

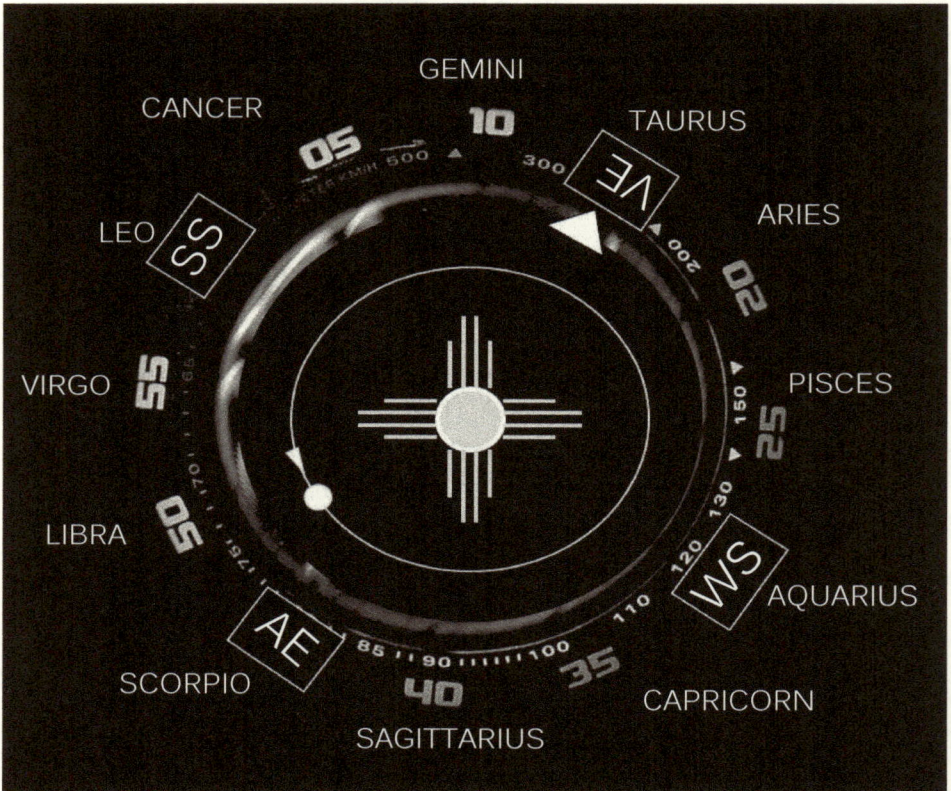

Using a watch-dial model to illustrate precession of the equinoxes. Imagine an aviator watch or a dive watch with a rotating outer bezel, with the main indicator on the dial marked by the triangle marked VE for "vernal equinox." As the earth rotates around the sun (small white circle on orbit around sun in the center), observers can look across to see that the sun will be rising in the constellation of Taurus on the spring equinox in this age. Other stations on the bezel are marked SS ("summer solstice"), AE ("autumnal equinox"), and WS ("winter solstice").

AGE OF ARIES

GEMINI
05
CANCER
SS
TAURUS
10
TER KMH 500
300
LEO
SS
VE
ARIES
200
VIRGO
50
20 PISCES
150
LIBRA
AE
25
AQUARIUS
130
85 90 100 110 120
40 WS
SCORPIO
35
CAPRICORN
SAGITTARIUS

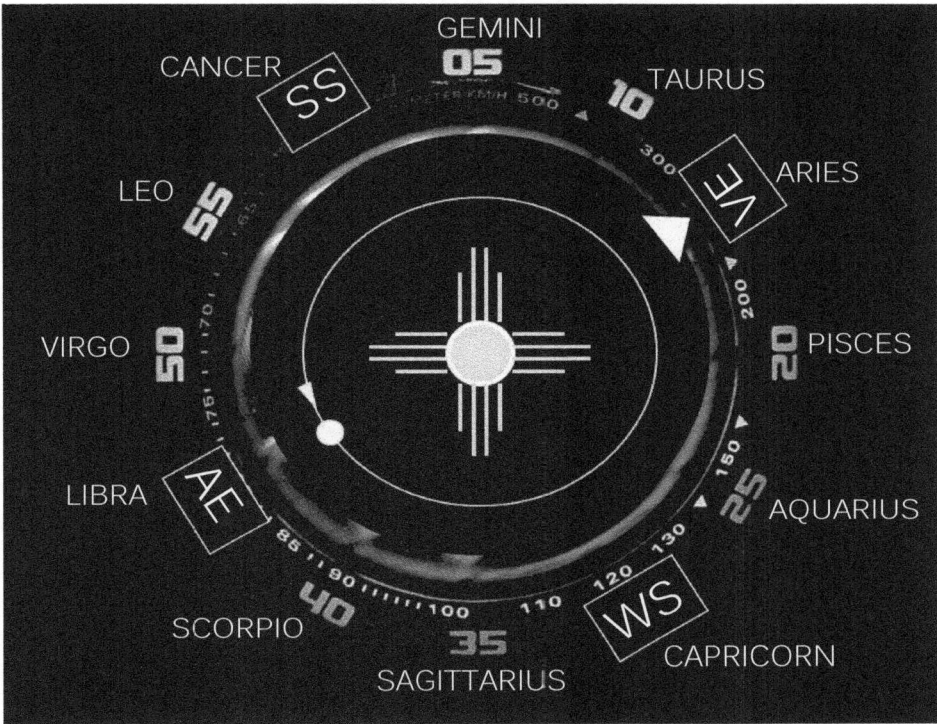

The Age of Aries. The vernal equinox (VE) is now in Aries. Summer solstice (SS) is in Cancer, autumnal equinox (AE) is in Libra, and winter solstice (WS) is in Capricorn. Remember on this special watch, the earth indicator (the small white disc) goes around the sun and will be positioned opposite the constellation in which the stylized sun in the center of the dial will rise on any day of the year. In the above dial, for example, the earth disc has made its way to the vernal equinox position of its orbit. When a point on the earth turns around and the sun begins to rise over the leading edge of the earth, the constellation all the way across the dial indicates the stars that will be in the sky just before the sun rises. In the Age of Aries, the constellation Aries will be in the sky just before the sun rises on the vernal equinox. The watch-bezel analogy helps illustrate the way each age has a group of four zodiac constellations associated with the four principal stations of the year (the two solstices and the two equinoxes). When the bezel shifts clockwise to the next age, the vernal equinox will shift but both solstices and the autumnal equinox will shift right along with it.

AGE OF PISCES

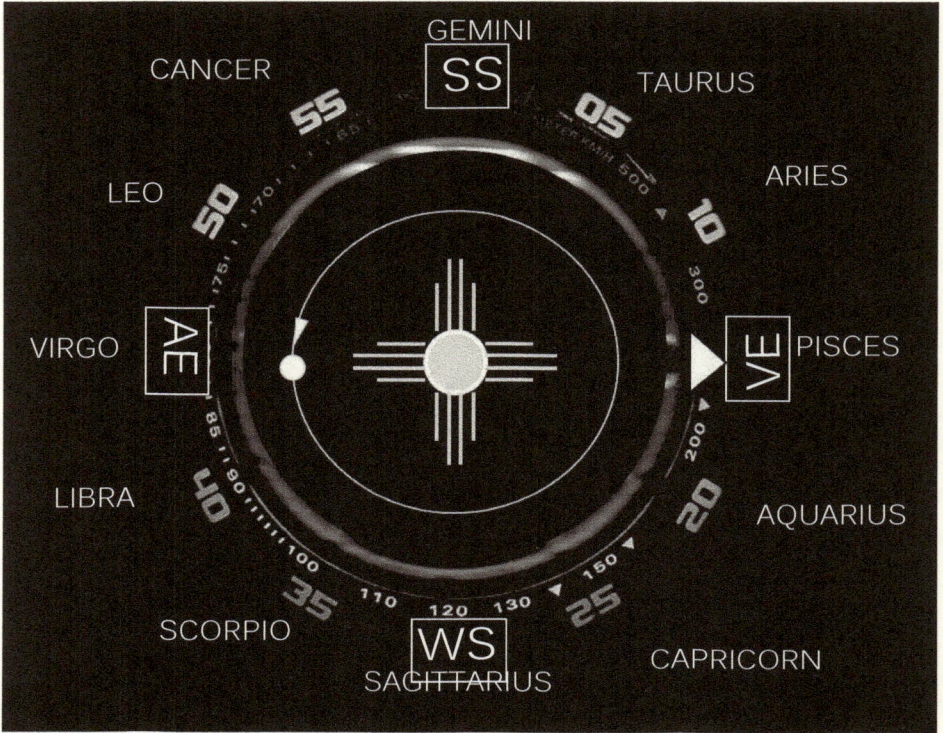

AGE OF AQUARIUS

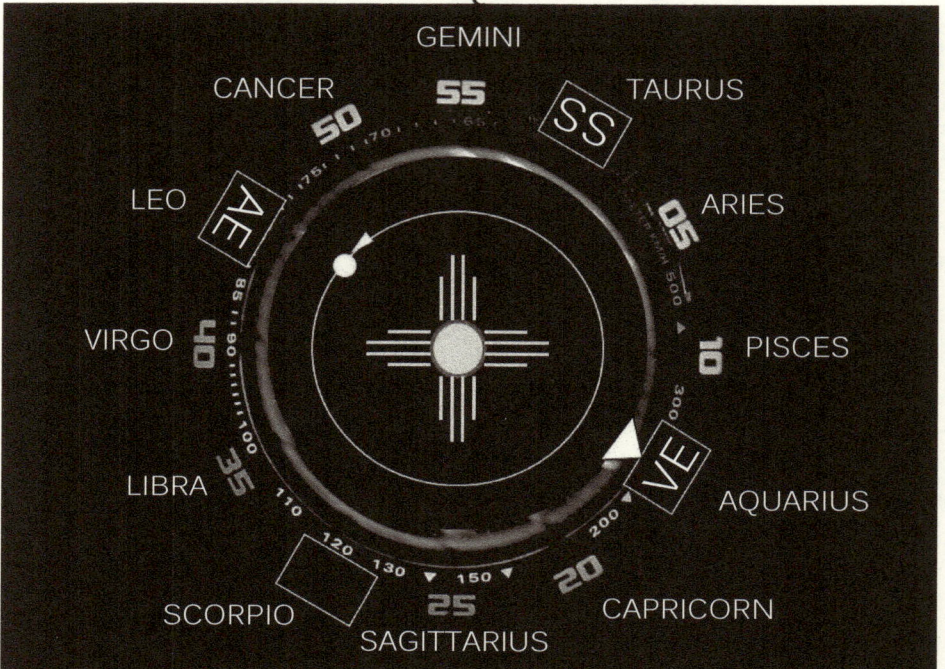

(AE), and the winter solstice (WS). The vernal equinox is addition-ally marked with a triangular indicator, since the vernal equinox traditionally began the annual calendar, and since the age takes its name from the constellation in which the sun appears to rise on the vernal equinox.

At this point in our discussion it is time to introduce the specifically astronomical term of the *colure*, apparently derived from a Greek compound word meaning "dock-tailed" or "stunted tail" ("so named because a part is always below the horizon" according to Webster's 1913 dictionary).

The colures are great circles running through the celestial pole, that central point in the night sky (and the daytime sky as well, although it is more difficult to visualize) around which the entire sky appears to turn – due, as we've seen, to the rotation of the earth on its axis. They are like enormous, imaginary hoops which frame the vault of the heavens, and tie together the motions we have been discussing. There are two colures, perpendicular to one another, intersecting at the celestial north pole and (below the horizon to those in the northern hemisphere) again at the celestial south pole.

One of these great hoops is defined by the celestial north and south poles and the two equinoxes, and the other by the celes-tial north and south poles and the two solstices. Hence they are termed the "equinoctial colure" and the "solstitial colure" (see the illustration on the following page).

To determine exactly where these two equinoxes and solstices are so that we can envision our two colures, we need one more hoop which is not called a colure (since it does not pass through the celestial poles, as a colure, by definition, does) but which is equally important, and that is the hoop formed by the travel of the sun (and all the zodiac constellations, which are, by definition, the constellations grouped along that line of travel of the sun) and which is known as "the ecliptic."

This circle of the ecliptic could be seen each day if you could attach an ink pen to the sun as it went through the daytime sky, leaving a line (and if, by some special property of the ink, that line could glow dimly after nightfall, you could see it through the night as well). That arc could be imagined in your mind as continuing below the horizon to form a complete circle.

What you are actually doing as you look along that arc in the sky is looking down the plane of the solar system, which makes sense if you think about what you are actually seeing when you watch the sun apparently travel across the sky.

When you were a child and you learned about the vast disk of the solar system with all its planets, you may have envisioned it as being on the same plane as the ground you are standing on, extending out from the circle of the horizon all around you. However, the way it should really be depicted is as though it were aligned along a great flat plane of glass which sticks out of the

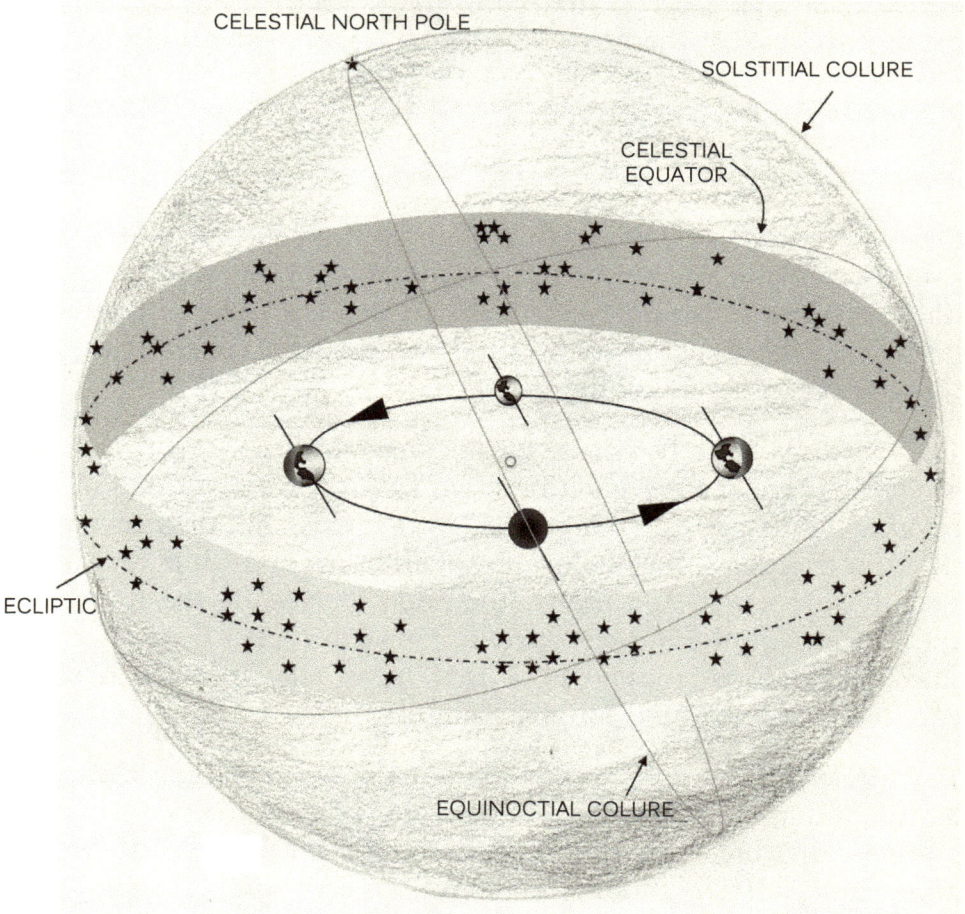

ground you are standing on at a degree of inclination roughly equivalent to the elevation of the sun as it traverses the sky during the day in your latitude of the world.

Thus, the ecliptic is the plane formed by the ellipse of earth's orbit around the sun, which to an observer on earth takes the form of the imaginary "plate of glass" we have been describing, which forms that incredibly important pathway through the sky along which the sun travels. It will change its angle through the sky, describing a steeper or flatter arc, depending upon the location of the earth on its orbit (due to earth's axial tilt). The reason this plane is called the ecliptic is that when the moon's plane intersects the plane of the ecliptic near a full moon or a new moon, it will produce an eclipse (the plane of the moon's orbit is inclined slightly to the ecliptic, or else the moon would produce an eclipse every month on the full moon and every month on the new moon).

We have already shown why the two points at which the celestial equator intersects with the ecliptic define the equinoxes.

So, we now have conceptualized all the important hoops framing the celestial sphere: the celestial equator, the ecliptic, the equinoctial colure (running through the north and south celestial poles and the two points where the celestial equator and ecliptic intersect, which are the equinoxes), and the solstitial colure (running through the north and south celestial poles and the two points where the celestial equator and ecliptic are farthest apart, which are the solstices).

We can now envision the mighty cogs which turn the critical points on the celestial sphere we have just framed with the hoops of the colures, ecliptic, and celestial equator. Remembering that all these hoops turn as one interconnected framework (which the authors of *Hamlet's Mill* sometimes call an "implex") we can imagine them turning through the constellations of the zodiac in each of the ages we've just described.

At left we see the earth going around the sun, with the ecliptic and the zodiac constellations along the zodiac lying on the plane created by the earth's orbit (the plane of the ecliptic, indicated by a dashed line). We see the celestial north and south poles, created by the turning of the earth on its axis. We see the celestial equator, 90° "down" from the celestial north pole (or 90° "up" from the celestial south pole). We see the equinoxes, marked by the point at which the celestial equator strikes the ecliptic. And we see the equinoctial colure, a circle defined by the north and south celestial poles and the intersections of the ecliptic and celestial equator.

Most importantly, we see the constellations of the zodiac which are opposite the sun along the line drawn from the earth to the

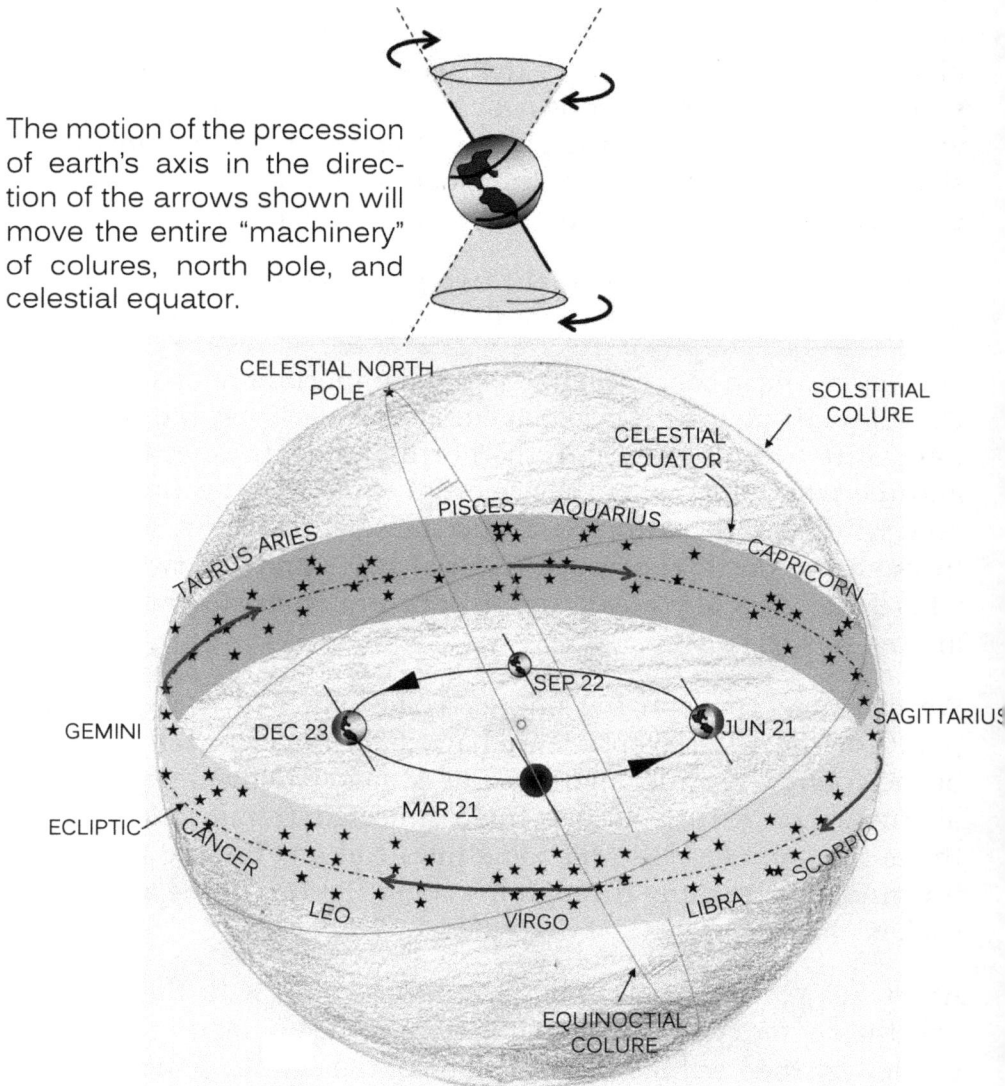

The motion of the precession of earth's axis in the direction of the arrows shown will move the entire "machinery" of colures, north pole, and celestial equator.

sun at each of the four important orbital stations – spring equinox (Pisces), summer solstice (Gemini), fall equinox (Virgo), and winter solstice (Sagittarius).

Now, in our minds, we can imagine the centuries-long gyration of the precession in earth's axis. Because this motion will actually move the celestial poles, the equinoctial colure which is defined by those celestial poles will actually shift – and because the solstitial colure is perpendicular to the equinoctial colure, they are basically "attached" and will shift together. And because the celestial equator will shift along with the celestial pole when the axis wobbles in its 25,800-year cycle, it will shift right along with them, dragging the equinoxes as it goes.

The only thing that doesn't rotate is the ecliptic, which – because the axis stays at its 23.5° angle to the earth's orbital plane around the sun (disregarding nutation) – is not impacted by the wobble which causes precession, since precession concerns the attitude of earth's axis, not its orbit. In other words, the ecliptic is defined by the ellipse drawn by earth as it orbits, and since the wobble of the poles does not alter this ellipse, the plane of the ecliptic remains while the colures and the celestial equator and the celestial poles themselves all shift along it.

In the diagram at left, we have added an arrow showing the direction of the precessional funnel drawn by the axis of the earth, and if we think about the axis gyrating in that direction, we can see that the colures and celestial equator must shift that same way – and now we see that the equinoxes and solstices are shifted by the rotation of this entire mechanism along the unshifting ecliptic, such that the colures will stand in the position of "the Age of Aquarius" – Aquarius (spring equinox), Taurus (summer solstice), Leo (fall equinox), and Scorpio (winter solstice).

Once the mechanics are understood, it is easy enough to envision turning the colures backwards to the previous age – the Age of Aries – and then to turn them back once more to the age which preceded that one, the Age of Taurus.

The important thing to notice here is that all the colures and circles shift together as one unified mechanism, except for the ecliptic and its zodiac constellations, which can be thought of as the cog ring around which all the others turn but which itself remains unchanged.

These mechanical metaphors, of cogs and watch bezels and clock parts, were not the metaphors chosen by the ancients who observed this vast but connected mechanism. Instead, they envisioned it using a variety of metaphors (actually, a dizzying variety, as Giorgio de Santillana and Hertha von Dechend detail in their stunning 1969 essay *Hamlet's Mill*).

One of these is the metaphor of the "World Tree," familiar from Norse mythology as Yggdrasil, whose branches and roots penetrate all the nine worlds. If you look up at the night sky and imagine the line between you and the celestial north pole as the trunk of a great tree, then all the stars and constellations might be leaves connected to this central hub, just as they appear to be in the sky. If the central pillar of the trunk were to become unseated, and begin to slowly rotate or gyrate, then it stands to reason that everything else, being connected and dependent thereon, would have to shift too.

This would appear to be a concept of staggering impact upon the minds of anyone who has gained familiarity with the regular and precise motions which, to all but the most scrupulous observer, seem to follow an identical pattern, year-in and year-out. What does it mean when the entire heavens, which appeared to glide unaltered along a track fixed for eternity, are suddenly revealed to be unhinged? And here we come to a profound point of interest: how is it that this mechanism of precision was known by virtually every culture of antiquity, whether in Asia, Europe, Africa, North America, South America, and even the remotest islands of the Pacific?

For not only was it known, it was known intimately and to a degree of advanced mathematical precision, and what is even more fascinating is that the details of these celestial mechanics appear to have been encoded in mythology that was connected to religious worship all over the globe.

Are we to believe, as the conventional theory of anthropology and human culture asserts, that the intricate and arcane concept of precession – an astronomical event measured in cycles of time so vast that one three-hundred-sixtieth of the entire event is as long as an exceptional human lifetime – was discovered by painstaking observations and measurements over the course of many generations, simultaneously around the globe by all these incredibly different cultures? Recall that the conventional historians do not even believe that precession was even discovered by the time these myths were composed.

Having discovered it, even though isolated from one another, are we then to believe that they all responded to it in a nearly identical fashion? Are we to understand that each of these widely dispersed cultures and peoples responded with a feeling related to religous worship? Are we then to understand that, having all decided that these subtle celestial mechanics should form the basis for religious worship, they all recorded their knowledge of this astronomical process in a "grammar" of architectural and mythological symbolism so strikingly similar that, when compared even across vast distances in time and place on the earth's surface, their similarities would be undeniable? Because the similarities, as we will see and as others have argued, are so plain and so abundant that they should cause any unbiased observer to immediately conclude that all were either descended from the same original knowledge-givers or that all of them were actually in communication in some way unknown to modern academia?

As ridiculous as these questions appear, conventional academia argues for an affirmative answer to all of them, and ridicules any who suggest the truth could turn out to be otherwise.

I believe there is a better explanation, and one which accords with the observed facts of geology, mythology, and archaeology far better than the conventional framework does (in fact, to say that the conventional theory accords at all with geology and mythology is, as we will see, overly generous).

For one thing, if (as conventional academia asserts) all of these cultures had observed the heavens for generations, learning their patterns and regularities, and then slowly realized that all those patterns were skewed by the ages-long process of precession, then why would they universally create myths with precessional imagery in conjunction with a myth about an event in which the axle of the sky was violently wrenched from its place?

What if, instead, a violent dislocation of the sky actually took place within human memory, and was recorded as a great dislocation of the mechanism of the heavens? This is in fact what the mythologies of culture after culture record as having happened. Is it possible that, instead of reflecting a postulated event that "must have happened" in the distant past and presenting it as fact, in a made-up story, these vastly dispersed cultures are recording a fact of which men had direct experience? Does it make sense to postulate that every single one of these vastly different and geographically remote cultures felt the identical need to make up a story about an event that "must have happened" at some point? Why would they all depict it that way? Why wouldn't some of them depict it as "the way things probably stood from the beginning?"

The main objection to such a logical suggestion – the suggestion that members of the human race experienced a catastrophic event that altered the sky, and passed down their knowledge to their descendents so that all these different cultures did not have to uncover by the painstaking process observation a motion that changes only 1° every 72 years – is the framework of history ushered in to support the theories of Darwinian natural selection. This theory requires vast eons of time – in fact, as the complexities of organisms were appreciated in greater detail since the time of Darwin, the amount of time demanded by the subscribers to this theory has expanded exponentially.

This Darwinian necessity has a direct impact on the timeline governing conventional theories of geology (the geological record must be interpreted through a lens that will support the assumption of incredible, almost inconceivable, age) and conventional theories of anthropology (mankind must be a relative newcomer to the stage). Under Darwinian assumptions, advanced human beings were not and could not have been present during the imagined eons of emptiness during which the forces which created the current heavenly patterns must have occurred, or during which sediments were laid down upon the earth's surface, or during the time that the Grand Canyon was carved, or the time that the Antarctic continent went from supporting lush vegetation – as the archaeological record there suggests it once did – to being covered in ice).

Those conventional frameworks are incorrect, and there is plenty of evidence that should cause everyone to consider new theories. However, the devotion of the acolytes charged with the defense and the advancement of the conventional picture of geology, anthropology and biology is ferocious. Most of all, they cling tenaciously to the Darwinian explanation of man's origins, and anyone who dares question the supporting assumptions required by the Darwinian theory must be prepared for withering criticism, ostracism, and marginalization.

Unfortunately for them, the facts of geology, archaeology, and what we know about ancient mythology – worldwide – are better explained by a fairly recent catastrophe, one which created the geological features we see on earth today, and which explains the evidence of seafaring, astronomically sophisticated ancients found in virtually every part of the world for those who care to look.

Armed with an understanding of the celestial mechanics set out in this chapter, we can now examine the evidence in mythology and archaeology which argues for a very different understanding of mankind's ancient past.

Evidence from myth:

Sumer, Babylon, and the Gilgamesh series

The celestial mechanics laid out in the preceding chapter are suffi-ciently complex that the knowledge of them has been forgotten many times even in recorded history. Even in our modern age, many are unaware of them. When and how did mankind first grasp the idea of precession, and why is it important to ask such a question?

The conventional framework of anthropology, wedded to the theo-ries of Darwinian evolution, has man groping forward through long ages of ignorance, slowly developing agricultural skills, and finally beginning to assemble the rudiments of scientific knowl-edge in the dawn of recorded history, particularly in ancient Egypt and more recognizably in ancient Greece. While this conventional view acknowledges the carefully recorded celestial observations of the Babylonians and Egyptians, it does not concede that preces-sion was understood or even clearly perceived before a relatively recent period, within two centuries before Christ, at a time when recorded "history" had certainly begun, and a time in which many of the actors on the stage of human events around the world are known to us through existing texts and archaeological evidence.

Only a few scholars, outside the comfortable boundaries of academic approbation, dare to suggest that the intricate machinery of the sky – and in particular the subtle phenomenon of precession – was grasped by man in some ancient era of which history knows nearly nothing, long before the inscription of the very earliest texts which survive today as lamps to illuminate (however dimly) our backward gaze.

While many of the arguments for an earlier comprehension of the mechanism of precession which follow have been put forth by other authors, they have not to my knowledge argued for the connection with the geological theories pioneered by West Point graduate and former MIT National Science Foundation Fellow Walt Brown, nor has Dr. Brown used this evidence from ancient anthropology to bolster his theory, although I believe the connections are compelling. I also hope to present some of the connections between ancient mythologies in a slightly more straightforward manner than they have perhaps been presented before, as well as draw one or two connections that perhaps have not been drawn previously.

The extensive if somewhat mysterious essay of Giorgio de Santillana (1902 – 1974) and Hertha von Dechend (1915 – 2001) entitled *Hamlet's Mill: an essay investigating the origins of human knowledge and its transmission through myth* (1969), is built around just such an incredible thesis: that some ancient civilization of which conventional history remains ignorant (perhaps willfully ignorant) unlocked these principles of celestial mechanics and encoded them in mythology, and that they then spread this mythology throughout the globe, to such a degree that remnants of the concepts they used can be definitively observed in the mythology of the pre-Columbian Aztecs of Central America, Cherokees of North America, Maoris of New Zealand, as well as in the more familiar mythology of ancient Greece which was already centuries old when it was included in the poetry of Homer and the discourses of Plato.

Even the mere detection of the slow precessional shift in the heavens requires the ability to accurately describe the positions of stars in order to record and track them over long periods of time, generally exceeding a human lifespan. It requires precise observational techniques and the ability to measure angles. It also requires mathematical tools, including much that we understand as belonging to the body of knowledge of trigonometry. Once detected, even more time and analysis is required before such a process is understood, and understood with a degree of precision and breadth that enables this celestial process to be described and predicted and passed on to others.

To suggest that such knowledge was known and codified long before recorded history, and only later re-discovered, is heresy to the world of modern academia. Indeed, the thesis of de Santillana and von Dechend has been deemed heretical by the guardians of modern intellectual orthodoxy, marginalized and ridiculed in order to discourage further serious analysis of their work or the subjects it treats.

The thesis of *Hamlet's Mill*, in the authors' own memorable words, is that:

> The dust of centuries had settled upon the remains of this great world-wide archaic construction when the Greeks came upon the scene. Yet something of it survived in traditional rites, in myths and fairy tales no longer understood. Taken verbally, it matured the bloody cults intended to procure fertility, based on the belief in a dark universal force of an ambivalent nature, which seems now to monopolize our interest. Yet its original themes could flash out again, preserved almost intact, in the later thought of the Pythagoreans and of Plato.
>
> But they are tantalizing fragments of a lost whole. They make one think of those "mist landscapes" of which Chinese painters are masters, which show here a rock, here a gable, there the tip of a tree, and leave the rest to imagination. Even when the code shall have yielded, when the techniques shall be known, we cannot expect to gauge the thought of those remote ancestors of ours, wrapped as it is in its symbols.
>
> Their words are no more heard again
> Through lapse of many ages. . .   4-5.

To suggest that such knowledge was dispersed beyond the oceans to the inhabitants of North America, Central America, South America and beyond to the islands of the South Pacific, before Columbus and his successors brought the technological accomplishments of civilizations intellectually descended from ancient

Greece, is even more blasphemous. And yet this is precisely what the clues lead us to believe.

The accepted view among scholars is that the phenomenon of precession was unknown until the time of Hipparchus of Nicaea (c190 BC – c120 BC), one of the most accomplished of the ancient Greek astronomers and an important predecessor of Ptolemy (AD 90 – AD 168).

Most of the works of Hipparchus are no longer extant. The only surviving text penned by Hipparchus himself is a critical commentary on the astronomy contained in the poem *Phaenomena* by Aratus of Soli, who died about fifty years before Hipparchus was born.

The word "phenomena," which has assumed a somewhat generic meaning in our language, appears to have been used by the ancient Greeks specifically in reference to the awesome celestial mechanics we have been discussing in the previous chapter, including the rising of the sun throughout the year and the turning of the band of the zodiac in the night sky.

The poetic *Phaenomena* of Aratus was based (according to Hipparchus, who accuses him of altering it for the worse in several cases) upon an earlier prose *Phaenomena* by Eudoxus of Cnidus, a contemporary of Plato who lived between 410 BC and 347 BC, almost a century prior to Aratus. None of the actual texts of Eudoxus have survived.

It is noteworthy that Aratus begins his poem, after his invocation to Zeus, by declaring that the stars "are drawn across the heavens always through all time continually [piling up adverbs to express their ceaseless wheeling] but the Axis [in notable contrast] shifts not a whit, but unchanging is forever fixed, and in the midst it holds the earth in equipoise, and wheels the heaven itself around" (lines 19-24, translated by A.W. Mair and G.R. Mair, Loeb Classical Library, London: 1921).

Perhaps this confident assertion of the immutability of the Axis of Heaven on the part of Aratus was the gauntlet that first aroused the crushing weight of Hipparchus' criticism; in any event, it serves to demonstrate the central importance of this "unmoved mover" around which all the heavers revolved but which itself "shifts not a whit, but unchanging is forever fixed." It is known that the *Phaenomena* was immensely popular and well-regarded, and was still so respected three centuries later that young Cicero spent much effort translating it into Latin.

From his commentary on the poem of Aratus, we can deduce much about Hipparchus' understanding of the heavens, which he brought to bear on the work of Aratus. Even further, we can observe the impact of his thought in references to his work in surviving texts, especially Ptolemy's "Great Treatise," the *Almagest*.

Otto Neugebauer (1899 – 1990), one of the most important modern scholars of the story of astronomy from Babylon to the time of Isaac Newton, wrote in his definitive three-volume *History of Ancient Mathematical Astronomy* (1975) that "Hipparchus' most famous achievement is unquestionably the discovery of precession" (292). Neugebauer explains that while "we do not know in what order Hipparchus' work proceeded," we do know that his accomplishments were gigantic. He compiled the first known catalog of stars, the only predecessor to the second known catalog, that of Ptolemy.

Neugebauer, following the work of Irish astronomer Robert Stawell Boll (1840 – 1913), believed that Hipparchus' catalog contained the positions of no more than 850 individual stars (fewer than the 1,022 stars of Ptolemy's catalog in the *Almagest*), and that based on the surviving details in the commentary on Aratus, Ptolemy's star location details were arrived at independently from those of Hipparchus, rather than being simple derivations of them as is commonly believed (285). Neugebauer also provides evidence that Hipparchus was in possession of and familiar with the extensive records of the earlier Babylonian astronomers, who had compiled copious arithmetical tables of the intervals between various celestial events.

We know that Hipparchus compared the positions of the stars he observed with the observations available to him from previous astronomers, particularly Timocharis of Alexandria (320 BC – 260 BC) and his contemporary and collaborator Aristyllus, and found that the star Spica (the brightest star in Virgo) had moved relative to the autumnal equinox. This was one of two observations which could only be explained by the phenomenon of precession (as Hipparchus had already definitively determined that the "fixed stars" – in contrast to the planets – did not wander but retained their relative positions to one another). As Neugebauer points out, due to the lack of surviving texts by Hipparchus himself, we do not know which observation came first or exactly what steps Hipparchus went through in his mind as he arrived at his "most famous achievement."

The second observation which reveals the fact that the great axis of heaven is in fact moving, contrary to the assertions of Aratus, was Hipparchus' analysis that, based on his measurement of the length of the tropical year (the interval between the sun's rising on successive equinoxes, a measurement of the year which only a few astronomers in that epoch even used as a way of reckoning time) differed by a slight but perceptible quantity of time from the length of the sidereal year (the interval between the return of a star to its rising point as it moves throughout the year – caused by the earth's annual orbit around the sun as discussed in the previous chapter). In other words, the turning of the zodiac, which makes a full rotation in a sidereal year as the earth returns to its same point in its orbit, was a tiny bit "off" from the steady movement of the sun as it marches between rising points of the solstices, passing the equinoctial points twice each year on its sentry-like circuit. This discrepancy, like the slippage of the star Spica, was also an undeniable indicator that the celestial mechanism was not unchanging throughout eternity.

Based on his observations, Hipparchus concluded that the equinoctial points were moving not less than $1°$ per 100 years in a direction opposite to the rotation of the zodiac.

The actual rate of shift (known as the "constant of precession"), as discussed in the previous chapter, is closer to $1°$ per 72 years. Neugebauer in his work discusses reasons to believe that Hipparchus may have actually calculated a closer estimate of $1°$ every 77 years, and that his parameter of $1°$ per 100 years was only meant as a lower limit. Ptolemy followed this lower boundary of Hipparchus in his own writings, and such was the high regard accorded to the *Almagest* and its systematic explanation of the heavenly phenomena that the constant of precession – when it was even acknowledged – was cited as being 1° per 100 years for the next several centuries, until the work of Muslim astronomers in the 9th century who began making their own observations and calculations and improved upon the constant set by Ptolemy.

These details reveal how arcane the knowledge of precession remained, even after the dawn of recorded history and the heights to which science and knowledge ascended under the greatest thinkers of Greece and Rome.   Nevertheless, there is evidence in the ancient myths of the world, most of which can be definitively placed on the timeline of human history centuries before Hipparchus and Ptolemy (centuries before Aratus and Eudoxus and Plato as well), which reveals a sophisticated understanding of the phenomenon of precession and its impact on the entire sphere of heaven, an understanding far more precise than that achieved by Hipparchus or Ptolemy or those who came after them for many centuries.

This assertion is not meant to detract from the towering achievements of Hipparchus and Ptolemy – far from it. Laboring in a field in which they themselves had to invent much of the terminology required to even mentally frame the concepts they were grasping with the powers of their focused intellect, they brought back into the light a field of knowledge upon which the shadows of ignorance had long ago descended. But, as Santillana and von Dechend assert, they were only re-discovering a science which had clearly existed at some point in man's distant past. From Hipparchus' and Ptolemy's vantage point in history, however, the effort required to re-discover it was every bit as heroic as that required to apprehend it in the first place, perhaps even greater, since all those who had once known it had long ago faded into dust themselves.

The authors of Hamlet's Mill are somewhat coy in their assertions. If the framework of high astronomical knowledge they perceive in the ancient mythologies were an elaborate mechanism in a darkened factory, then Santillana and von Dechend in their massive work prefer to approach it with flashlights, shining a little here and there before moving elsewhere, never fully illuminating the entire machine, but revealing enough to convince the observer that there is certainly *something* there.

Part of the reason for this methodology is their displeasure with those who are always ready to put mythology into a box, over-simplifying the tangled clues left to us from the ancients and trying to reduce all the deep mystery, to explain away all of the complications, contradictions, and deliberate disinformation that we find and pretend that all is clear and uncomplicated.

Another reason they do not reveal the entire masterpiece is that it is no longer complete – they could not reveal it in its entirety even if they wanted to. De Santillana and von Dechend compare the body of knowledge they treat in their essay to Bach's lost series of compositions known as the *Art of the Fugue*. They write:

> Bach's *Art of the Fugue* was never completed. Its existing symmetries serve only as a hint of what might have been, and the work is not even as Bach left it. The engraved plates were lost and partly destroyed. Then, collected once more, they were placed in approximate order. Even so, looking at the creation as it now is, one is compelled to believe that there was a time when the plan as a whole lived in Bach's mind. 346.

In the following chapters, we will attempt to trace some of the most complete strands of mythological evidence offered in *Hamlet's Mill*, and apply that analysis to some of the related evidence from ancient archaeological sites that *Hamlet's Mill* does not cover (its authors being primarily concerned with the evidence from myth). We will also examine some of the more astonishing pieces of evidence and analysis that have come to light since *Hamlet's Mill* was published in 1969.

The first piece of the puzzle to pick up might logically be the string of mythological artifacts surrounding the axis of heaven, which as we have seen in the preceding chapter on celestial phenomena is not at all an axis "unchanging forever fixed," shifting "not a whit" as described by Aratus of Soli.

To the contrary, the axis of heaven – which is a projection of the earth's actual axis, tracing its precessional circle in space – is slowly moving, replacing the celestial north pole over the centuries, and with it moving the entire framework of heaven: the celestial equator, the equinoctial and solstitial colures, and therefore the equinoxes and solstices themselves.

What Santillana and von Dechend discover is that this concept of the shifting axis is found throughout ancient mythology, from the ancient Greek myths that were centuries old before Homer's time, to the ancient Hindu Vedas dated by conventional scholars to 1100 BC or even earlier, to the Babylonian epic of Gilgamesh dated to 2000 BC or even earlier, to echoes found in the mythologies of Asia, the Americas, Oceania and Africa.

The axis analogy in these myths often takes the form of a god standing on one leg, or a god who replaces another god who retires from the scene, or a combination of both of these (a young god who stands on one leg and replaces the previous god who stood upon one leg), or a central "world tree" that is uprooted or cut down or replaced, or the hub of a millstone that spins off of its axis and into the sea, or a conical mountain (often inverted such that the more pointed end is the base of the mountain and the top is broader), or a bent central pole that whirls and is associated with producing fire.

Often, the imagery involves a cube that is located at the "navel of the world," or a nail or a plug at the center of the world, the removal of which is associated with a catastrophic flood, or with rising water that threatens a catastrophic flood. The connection between the dislodging of the central axis of the universe and the sea or the flood is found across many cultures and mythologies, and may be significant for reasons that fit together with man's ancient past, as we shall explore below.

Among the most ancient texts available for examination by modern scholarship are inscribed on the tablets of what we call today the *Epic of Gilgamesh*. Many scholars, in fact, believe the ancient Sumerian and later Babylonian texts among which the series of Gilgamesh were discovered to be the most ancient surviving written works of literature we possess.

They were lost and forgotten for over two thousand years, lying among the ruins of the ancient Assyrian capital of Ninevah beneath modern-day Mosul in Iraq until the 1840s and 1850s, when the library of Asshurbanipal the last king of Assyria (who lived from 668 BC – 627 BC) was discovered among the buried ruins of his palace.

Beginning in 1853, English archaeologists sent over 25,000 clay tablets inscribed with cuneiform writing – the earliest known writing system in the world, still in the process of being deci-phered at the time – back to the British Museum in England. It wasn't until 1872 that the Gilgamesh texts were finally noticed among the tablets, when a young curator of the British Museum recognized with tremendous excitement the account of a "Baby-lonian Noah" who had survived a great deluge, whose ship came to rest upon the mountains of Nizir, and who sent out a dove that returned after finding no resting place.

The oldest tablets involving Gilgamesh, recovered from the library of Asshurbanipal, were Sumerian, written in verse and in Sume-rian, an ancient language with no known related languages. As A. R. George, Professor of Babylonian at the University of London's School of Oriental and African Studies, explains in the introduc-tion to his critical edition of 2003 (containing translations of all the known extant Gilgamesh texts), there are some scholars who have alleged finding references to Gilgamesh in the Old Sumerian tablets dated to as early as 2600 BC. However, based on his own examination of these tablets, Professor George found no grounds for specific identification of Gilgamesh in these texts. His own conservative estimate is that:

> The early rulers of Uruk had a great impact on the poets of the third millennium [3000 BC to 2001 BC], much as the Trojan war and its aftermath had on Homer. The reigns of Enmerkar, Lugalbanda, and Gilgameš entered legend as the heroic age of Sumer. One can imagine that court minstrels and storytellers began to compose oral 'lays of ancient Uruk' soon after the lifetime of these heroes, and it would then be no surprise for epic tales of Gilgameš and his predecessors in due course to appear in writing. At the moment one cannot

be sure that this happened in the Early Dynastic period, but it had certainly happened by the end of the millennium. 6.

After the rise of Babylon, where Akkadian was spoken, the Old Sumerian texts apparently became favorites for scribal translation and copying into Akkadian (still using cuneiform on clay). George explains that by the eighteenth century BC (1800 BC to 1701 BC), Old Sumerian was no longer in use in places in which poems and epics were recited for entertainment, and was mainly lost outside of the halls where those in training to become scribes would be schooled. "The number of Old Babylonian manuscripts extant for any given composition of the Sumerian corpus thus reflects its popularity as a school copy-book. Bilgames and Huwawa A, the story of Gilgameš's expedition to the Cedar Forest, was by far and away the most popular of the five poems, a fact that is explained by the recent discovery that it constituted the last of ten compositions in the second group of set texts encountered by the would-be scribe" (8).

Thus, we can conservatively date the Old Sumerian Gilgamesh texts as being composed no later than 2000 BC (and perhaps a century or more earlier), and the Old Babylonian Gilgamesh texts as dating to 1700 BC or 1800 BC, perhaps a little before. In other words, the Gilgamesh series is almost unbelievably ancient, from a civilization that existed over twenty centuries before the time of Christ – as far removed from the days of Rome and the first Caesars as we are, but in the other direction.

It would be shocking to find coded references to the sophisticated celestial machinery of precession in a record so ancient.

Yet when we turn to that poem which appears to have been a favorite in ancient Babylon and a required hurdle for all aspiring members of the scribal caste, the story of Gilgamesh and Enkidu's expedition to the Cedar Forest in search of the mighty Humbaba, we find clues which resonate with the precessional metaphors found in ancient myths the world over.

After the tablets describing the creation of Enkidu, the "double for Gilgamesh, his second self" in the new English version of Stephen

Mitchell (page 74), and the account of the taming of Enkidu and his becoming like a brother to Gilgamesh, Gilgamesh declares that he and Enkidu must travel to the Cedar Forest, where the fierce monster Humbaba lives.

Enkidu is stunned – he is familiar with the Cedar Forest, and he tries to dissuade Gilgamesh with the information that the Cedar Forest is endless. But Gilgamesh will not be dissuaded: "Listen, dear friend, even if the forest is endless, I have to enter it, climb its slopes, cut down a cedar that is tall enough to make a whirlwind as it falls to earth" (Mitchell, 92).

Is this just an epic version of mundane quest for commodities such as timber in the ancient Sumerian world? No. The celestial import of the quest is clear from the events which follow – the invitation and rejection of Ishtar, and the battle with the Bull of Heaven, which Ishtar sends to earth to destroy Gilgamesh and his city (after obtaining permission from her father, the god Anu) in revenge for being spurned. As the extensive studies cited by de Santillana and von Dechend demonstrate, what we might call the easy "high-school English-class essay" attempts to declare that "Gilgamesh stands for this" and "the cedar stands for that" are extremely ill-advised with a work of literature as rich and compli-cated as the Gilgamesh series, especially since it comes to us in numerous tablets full of lacunae or gaps, and full of critical words whose translations are still in dispute or completely unknown among existing scholars.

However, it is fairly clear that astronomical mechanisms are contained in the text, without taking anything away from the deep issues of human existence with which the stories also deal so powerfully. In fact, it would be possible to argue that the very power of the literary frame helped create an exquisite and unfor-gettable vessel for transmitting sophisticated and arcane celestial truths – a vessel so precious and beloved that it became a favorite among scribes for centuries to come.

The strange warning of Enkidu, that the forest goes on forever, is one of the first clues that we are dealing with a metaphorical forest that is no earthly forest at all. Further, as tablet V of the epic opens, and Gilgamesh and Enkidu "stand gazing at the forested slopes of the Cedar Mountain," Professor George informs us, "they perceive that the mountain is the residence of the gods and goddesses" (466). There is something unworldly about this forest, which we have already seen is an endless forest, and which blankets a steep-sided mountain that is the abode of the pantheon of gods. As George explains, the mountain of the gods has precedent in Sumerian tradition, described as "the mountain of both heaven and earth," but "this is a cosmic location, not a terrestrial one" (466, footnote).

He also notes that, according to the most ancient Sumerian texts, in their journey to reach this sacred mountain, the heroes had to cross seven mountain ranges guided by the constellations. Is it possible that theirs is actually a celestial journey, through the starry expanse itself, in search of the central tree that turns the entire celestial sphere? As an indication that this was to be no ordinary journey, we witness a scene in which, prior to embarking on their hazardous quest, the heroes seek divine anointing to protect them on their journey (in some versions visiting Gilgamesh's goddess mother Ninsun).

If the Cedar Forest has celestial import, what would the tallest cedar in the forest represent? And why would it be necessary to cut it down? It is clear from the events that follow that Humbaba, the guardian of the great cedar, was placed there to protect it by the god Enlil himself.

Gilgamesh battles with Humbaba, and after a struggle so ferocious that Cedar Mountain is split in two (in one Old Babylonian version, sundered by the deafening roar of Humbaba), Gilgamesh slays the monster with his enormous axe.

Then Enkidu declares, "By your great strength you have killed Humbaba, the forest's watchman. What could bring you dishonor now? We have chopped down the trees of the Cedar Forest, we have brought to earth the highest of the trees, the cedar whose top once pierced the sky. We will make it into a gigantic door, a hundred feet high and thirty feet wide, we will float it down the Euphrates to Enlil's temple in Nippur. No men shall go through it, but only the gods. May Enlil delight in it, may it be a joy to the people of Nippur" (Mitchell, 128 – 129).

In the next episode in the Gilgamesh series, Gilgamesh and Enkidu slay the Bull of Heaven, which is sent by Ishtar (Sumerian Inanna) when Gilgamesh resists her advances. After the hero slays it, he hurls its haunch in Ishtar's face (some versions of the text have Enkidu hurling the bull's haunch). The celestial significance of the bull itself (the constellation Taurus) as well as the bull's haunch (associated in ancient Egypt with the constellation we call the Big Dipper) are not disputed even by most conventional scholarly consensus.

The connection between the chopping down of the cedar, "whose top once pierced the sky," and the slaying of the Bull of Heaven, is made explicit in the ancient texts, when the god Enlil declares that for these two acts, one of the duo of Gilgamesh and Enkidu must die.

It is very possible that the cutting down of the tallest cedar in the celestial forest represents the unhinging of the very axis of heaven itself, the dislodging of the celestial north pole, setting loose the precession of the entire heavenly sphere. The description of the fall of the tree as one which induces a "whirlwind" should not be overlooked (indeed, if we return to the diagrams in the previous chapter, we see that the stars appear to "whirl" about the celestial equator, like a whirlpool – mentioned in many mythologies – or a whirlwind).

As if to be certain that its significance is not lost, the felling of this great tree is coupled with the slaying of the celestial bull – an event with clear precessional implications, as can be seen by the "watch dials" in the previous chapter depicting the two ages previous to the one we inhabit now. Two ages before this one, the Age of Taurus gave way to the Age of Aries, just as we are now on the verge of the end of the Age of Pisces and the inception of the Age of Aquarius. The slaying of the Bull of Heaven may well be connected to the precession of the vernal equinox out of the sign of Taurus and into the sign of Aries (an event of great significance echoed in other mythologies and cultural expressions, such as the "tauroctonies" of the Mithraea associated with the Mithraic mystery religion of the 1st century AD, to which we shall return later).

A third clue, albeit more arcane, lies in the fact that Gilgamesh declares his intention to fashion a door from the cedar he has cut down, a door through which only gods and not men may pass.

Oddly enough, doors appear to have an equinoctial function in myths worldwide. As we have seen, the equinoxes are the points where the ecliptic, that extremely important pathway of the sun and the planets, crosses the celestial equator, twice each year. Thus the crossing points can be analogized as portals, critical chokepoints, terrain which would be "key terrain" to a military tactician.

The authors of *Hamlet's Mill* identify several "clashing doors" in mythology around the world, the most familiar to most western readers being the Symplegades, the clashing rocks through which Jason and the Argonauts must pass with their ship.

In the tale of the Argonauts, Jason is told by Phineas to release a dove before attempting to pass between the clashing rocks; the dove darts through, losing only a few tail feathers that are caught between the rocks. Following its timing, the Argo shoots forward and reaches the other side; only the stern ornament of the ship is crushed between the snapping cliffs. Defeated, the Symplegades snap open and never crash together again: the analogy in the myth is that the Argonauts have opened a new door, a new pathway – the symbol of a new pair of equinoxes, a new channel through which the ecliptic can cross the celestial equator. The creation of a new door, then, is analogous to the advent of a new equinoctial passageway – the shifting of the equinoxes due to precession.

Thus, the felling of the cedar whose "top pierced heaven," a cedar whose felling is associated with a whirlwind, is associated with the creation of a "new door" through which gods (the sun and the planets) can pass. The celestial imagery is compelling. Add to these mythical metaphors the slaying of the Great Bull of Heaven, which is clearly Taurus, the guardian of the vernal equinox before the dawning of the Age of Aries, and it is difficult to argue that someone did not know about the concept of precession literally thousands of years before the birth of Hipparchus. The clues in one of the most ancient texts known to man are extensive.

But that is just the beginning of the trail of evidence.

Before returning to evidence from myths contemporary in time to the Gilgamesh epic, let us jump briefly to the "New World," where we find a record from the Omaha, a proud tribe of Plains Indians whose territory extended from Nebraska to South Dakota contiguous with the Sioux, with whom they were often at war.

Ethnologist Alice C. Fletcher of Harvard's Peabody Library lived among the Omaha in the 1880s and published several accounts of their cultural traditions and rituals. It is important to realize that she had access to first-hand accounts from Omaha elders and chiefs who had lived during the time prior to the abrupt disruption of their way of life caused by the sudden invasion of white settlers and in particular the rapid destruction of the buffalo herds beginning in the 1870s.

In *The Omaha Tribe*, published in the *27th Annual Report of the Bureau of American Ethnology to the Secretary of the Smithsonian Institution, 1905-1906* (published in 1911), she (along with co-author Francis La Flesche, a member of the Omaha Tribe) describes a cultic item of tremendous importance to the Omaha, "The Cedar Pole":

> An ancient cedar pole was also in the keeping of the We'zhiⁿshte gens, and was lodged in the Tent of War. This venerable object was once the central figure in rites that have been lost. In Creation myths the cedar tree is associated with the advent of the human race; other myths connect this tree with the thunder. The thunder birds were said to live "in a forest of cedars." The phenomenon of lightning striking a tree was explained as "the thunder bird has lit on the tree." What, if any, relation existed between the rites connected with the Cedar Pole and those of the Sacred Shell can not now be ascertained among the Omaha. The fact that both these relics of past ceremonials were in charge of one gens would seem to indicate some sort of connection. This Cedar Pole was called Waxthe'xe, a name afterward transferred to the Sacred Pole in charge of the Hoⁿ'ga gens

[this Sacred Pole is still in existence, having been transferred to Harvard's Peabody Museum in 1888 and returned to the Omaha people in 1989]. The Sacred Pole symbolized the power of the chiefs and it is not improbable that the Cedar Pole stood for the power of Thunder, the god of war. The Cedar Pole was 1 m. 25 cm. in length. To it was bound by a rope of sinew a similar piece of rounded cedar 61 cm. long called the *zhi'be*, or "leg." In the middle of the pole was bound another rounded piece of the wood, steadied by a third and smaller one, as three round sticks can be bound together more firmly than two. It is said that the pole typified a manlike being. As stated above, the lower piece was called "the leg," and it may be that the stick bound to the middle represented a club. One of the ritual songs used in the ceremony for awarding honors says:

> Behold how fearful is he, your Grandfather
>
> He lifts his long club, fearful is he.

There is a tradition that in olden times, in the spring after the first thunder had sounded, in the ceremony which then took place this Cedar Pole was painted, with rites similar to those observed when the Sacred Pole was painted and anointed at the great tribal festival held while on the buffalo hunt. If this tradition is true, these ceremonies must have taken place long ago, as no indication of any such painting remains on the Cedar Pole. 457-458.

The discovery of this Cedar Pole is recounted in several descriptions of the Omaha culture, and the account contains details that resonate with the description of the Cedar Forest and the felling of the tallest cedar in the Gilgamesh series.

As Dr. Edwin C. Krupp relates in *Echoes of the Ancient Skies: the Astronomies of Lost Civilizations* (1983):

> The Omaha Indians preserved a myth of a sacred cedar tree upon which the stability of their society was founded. During a time of tribal disruption, brought on by rivalries between the Omaha chiefs, one of their sons encountered a burning

cedar tree. Although enveloped in flames, the tree did not burn up. The chief's son noticed that the forest animals had worn four trails to the tree, one leading to each cardinal direction, and in the tree's branches roosted thunderbirds, which are emblems of the sky and the talismans of warriors. Upon learning of the burning cedar, the Omaha warriors dressed for battle and attacked it. They brought it down and transported it to their village. There the tree was reerected, and responsibility was assigned to one family. Through the tree, the once-threatened social order of the Omaha was strengthened. Problems, disagreements, and troubles were all brought – with presents and prayers – to the sacred pole. Leadership of the tribe was invested in the keepers of the pole, and only through them could authority over the tribe be transferred to others. 88.

The parallels to celestial events are fascinating. The sacred cedar tree is set apart by the presence of supernatural creatures – the thunderbirds – and by the remarkable symbology of burning without being consumed.

We are told that the tree is in the center of the four pathways that mark the four corners of the world – in celestial terms, this would mean not merely north, south, east and west but the four stations of the annual celestial cycle, the two solstices and the two equinoxes (which are the "pathways" of celestial animals as well, the zodiac animals).

We see that this cedar tree is somehow associated with strife and contention among the leaders (a theme that will become more and more prominent as we examine other myths concerning the Precession and its implications for heavenly rulership, such as the battle between the gods and titans in Greek mythology, which is echoed in other mythologies worldwide).

And we see that the ancient legend depicts a strange response to the discovery of this marvelous cedar – it is immediately chopped down! Why the urgent desire to topple this symbol of the celestial colures governing the solstices and equinoxes, a desire found apparently in both hemispheres among cultures that (according to conventional anthropology) have no historical contact? It would seem to symbolize the unhinging of the celestial pole, a pole that actually is unhinged by the precessional wobble of earth's axis, a wobble which disrupts the "four paths" of the solstices and equinoxes as well.

The connection of this pole with the celestial axis is more explicit from its connection with the Sacred Pole, mentioned above by Alice Fletcher. The Sacred Pole, which was symbolically painted red during the summer buffalo hunt, was accorded a place of honor in the midst of the Omaha encampment, and it was always inclined in a sort of crutch so that instead of pointing straight upwards, it inclined at about 45°, and was always positioned to point towards the north. In other words, at the latitude at which the Omaha ranged, it was always aligned in the ground to point to the celestial north pole! In this manner, it clearly became a physical manifestation of the invisible celestial axis running through the north pole. And let us not overlook the remarkable detail preserved by Alice Fletcher's account that this representation of the celestial axis had a small rounded stick called the "leg" bound to it by a sinew. We have already seen in Gilgamesh that the "haunch" of the bull is connected in ancient myth with the circumpolar constellation of the Big Dipper, and that this haunch is mentioned in close proximity to the cedar tree in the Gilgamesh epic as well. This should confirm that the cedar pole of the Omaha represents the celestial axis, as it has a "leg" tied to it, just as the one in the sky does.

Further, as Alice Fletcher explains elsewhere in her report to the Smithsonian Institute cited above, at the annual ceremony of anointing the Sacred Pole, twelve ceremonial verses were recited to it, six by women and the other six by men. The celestial import of the number twelve is quite clear worldwide. But why would cultures which never came into contact with one another select twelve as a number related to the ring of the heavens? The ancient sexigesimal system (based on sixes and sixties, which are factors of 360, the number into which near and middle eastern cultures including Babylon and Egypt divided the circle and the year) lends itself to a division of twelve, and we would perhaps expect to encounter celestial "twelves" in cultures that may have had contact with Babylon or Egypt, but is not remarkable to note the use of twelves in cultures such as the Omaha?

Again, perhaps this concept of a tree, representing the celestial phenomena of the celestial pole and solstitial and equinoctial axes, is so obvious that it is not surprising to find it represented in cultures as far dispersed as ancient Sumer, Babylon, and the Omaha of North America. That argument is the necessary hypothesis of the "isolationist" conventional anthropologists, who cannot countenance any contact across the oceans in ancient times not known to their recorded histories. These distant and unconnected cultures must have just independently hit upon the analogy of a Cedar Forest for the endless expanse of the heavens, and the chopping down of a celestial cedar for the unhinging of the sky, which they may have associated in some way with the initiation of the phenomenon of precession (which neither culture is supposed to have even understood, according to conventional views of their levels of astronomical understanding). Isn't it "natural" to analogize the celestial phenomena in terms of sacred cedar poles?

To anyone who goes out to the night sky and looks up, the "obviousness" of the tree symbology is not nearly so obvious. However, if these two cultural references to cedar trees and celestial symbols were the only two pieces of evidence for theories opposing isolationism, perhaps the conventional theorists could argue them away. Unfortunately for their thesis, however, the clues continue to pile up, and as we go along they will decrease in "obviousness" to the point that their similarity between peoples on different hemispheres who supposedly have no historical contact will become more and more startling.

Before leaving the cedars of ancient Sumer, ancient Babylon and the Omaha Nation of North America, we pause to point out that so far we are finding fairly compelling evidence that people knew of precession literally thousands of years before the conventional model currently admits. We are also finding more than passing evidence that the analogies that preserved this knowledge (such as the analogy of a cedar pole for the unhinged axis of heaven) were somehow transmitted across the oceans to the "New World" (or across the oceans the other way). Further evidence will make such conclusions harder and harder to deny.

Evidence from Egypt:

the myth of Osiris, Seth, and Horus

We have seen some hints of "remarkable coincidences" in the mythologies of far-flung cultures (far-flung both in terms of time – from ancient Sumeria to almost the present day – and in terms of actual geographical distance) that hint at very ancient cultural contact or even common origin. We now turn to the ancient myths of Egypt, famous even among writers in the ancient world for their awesome antiquity, for their centuries of enduring power over the Egyptian culture and surrounding cultures (including the Romans), for the religiosity of their followers (Herodotus famously saying there were none in the world so religious as the Egyptians – "They are religious to excess, beyond any other nation in the world") and for their daunting inscrutability and mystery (Herodotus, *Histories*, Selincourt, 1954, 143).

It is quite clear from the ancient Egyptian texts and scenes that there is a strong celestial connection to the divine events they relate and depict. The connection of the star Sirius (Sothis in Greek) with Isis and of the constellation Orion with Osiris (or the soul of Osiris) was explicit, and known to Greeks such as Herodotus who traveled to Egypt and described their religion and beliefs. It is also clear that the mysterious events and figures of the Egyptian pantheon had direct echoes in other pantheons, including those of ancient Greece, and the Greek writers often substitute the names of their deities when describing the Egyptian, for example Hermes for Thoth or Typhon for Set.

The importance of the ancient Egyptian mythology as a source of evidence for the extremely ancient awareness of precession, as well as the reverberation of characters, themes and events found in ancient Egyptian mythology in the mythologies of cultures across the globe (powerful evidence for some ancient common source) warrants treatment at some length.

We will focus here mainly on the evidence for connections with the celestial phenomena and for connections with other mythologies around the world. There are many fascinating aspects of the funerary rites, tomb layouts, and symbology in the Egyptian religion, some of which we will touch on. Its complexity and richness (layers upon layers of analogy and symbol, many of which remain a mystery to this day) warrant entire books, and the reader is encouraged to make use of the more thorough examinations found in the *Death of Gods in Ancient Egypt* (discussed below), *Hamlet's Mill* and the earlier sources referenced by its authors, and the very readable ancient sources such as the texts of Plutarch and others.

The civilization of ancient Egypt stretched literally across millennia, but some of the most ancient of texts dating to the Old Kingdom were discovered by modern man only in the 1880s, the most ancient currently known being the Pyramid Texts found inscribed upon the walls and sarcophagi of the Saqqara pyramids.

The Pyramid Texts known at this time include those from the tombs of:

> Unas (or Unis) (2353 BC – 2323 BC)
>
> Teti (2323 BC – 2291 BC)
>
> Pepi I (2289 BC – 2255 BC)
>
> Ankhesenpepi II, the wife of Pepi I
>
> Merenre (2255 BC – 2246 BC)
>
> Pepi II (2246 BC – 2152 BC)
>
> Neith, Iput II, and Wedjebetni, all wives of Pepi II
>
> Ibi (2109 BC – 2107 BC)

(James P. Allen and Peter Der Manuelian, *Ancient Egyptian Pyramid*

Texts, volume 2005, part 2, page 1; the authors use "circa" on all the above dates).

Assuming the above dates are accurate, this places the Pyramid Texts in the same millennium as the original Sumerian versions of the Gilgamesh series, although the exact century of the origin of the Gilgamesh poems is less precisely known.

As James P. Allen and Peter Der Manuelian explain in their introduction, the content of the Pyramid Texts is almost certainly older than the dates of the Pharaohs and their wives given above. Reflecting the state of scholarly consensus in 2005, they write:

> Although they are first attested in the pyramid of Unis, most of the Pyramid Texts are undoubtedly older. With few exceptions, their grammar is that of a stage of the language that disappeared from secular inscriptions at least fifty years earlier, and the architecture of the pyramid chambers that they reflect [. . .] came into use at the end of the Fourth Dynasty, more than a hundred years before Unis's time. Some of the texts also reflect burial practices that are even older, in earthen graves beneath tombs of mudwork. Newer spells that first appear in the later pyramids, however, incorporate features of the contemporary language. Overall, the Pyramid Texts give the impression of a corpus that had been in use for some time before it was inscribed in Unis's pyramid and one that was continually revised and amplified during the reigns of his successors. The process went on after the end of the Sixth Dynasty, in the Eighth-Dynasty corpus of Ibi and that prepared for the burial of the Ninth-Dynasty king Wahkare Khery (ca. 2030 BC). The Coffin Texts of the Middle Kingdom incorporate copies and revisions of some Pyramid Texts, and are mostly a continuation of the older tradition rather than a distinct corpus.

The thesis explained in the passage above, that the texts date to a time somewhat earlier than the reign of Unis or Unas, to the Fourth Dynasty, places them as early as the time that conventional opinion places the construction of the Great Pyramid and the pyramids of Khafre and Menkhare (the three pyramids most well known around the world, discussed in greater detail later in the section on clues from ancient archaeology), and potentially earlier than that.

Wood from the boat found near the Great Pyramid, the pyramid which conventional scholarship attributes to Khufu, has been carbon dated to about 2600 BC (see for instance the interview with archaeologist Mark Lehner of the Oriental Institute of the University of Chicago and the Harvard Semitic Museum online at   http://www.pbs.org/wgbh/nova/pyramid/explore/howold2. html).   This gives rise to the conventional dating of the Great Pyramid to around 2600 BC (since they attribute both the boat and the pyramid to Khufu's reign).   Whether the Great Pyramid dates to 2600 BC or whether it is even older, as some have maintained, the point for the discussion of the Pyramid Texts is that the Great Pyramid (which like the other two famous pyramids contains almost no inscriptions) is at least from that early date, and the scholarship mentioned above dates the content that we have preserved in those later Pyramid Texts to a similarly ancient date.

As Jane B. Sellers, author of *Death of Gods in Ancient Egypt: A Study of the Threshold of Myth and the Frame of Time*, has convincingly demonstrated, the Egyptian mythology referenced in the Pyramid Texts and other extremely early writings contains references to the Osiris myth which in itself appears to incorporate precessional imagery and understanding, long before Hipparchus.

Sellers has been adamant that she does not rely on the mention of precessional numbers as part of this evidence in the most ancient texts, for instance the fact that in the Osiris myth, Set or Seth is said to kill Osiris with the help of his gang of 72 helpers (72 being a clear precessional number, in that precession currently delays the rising point of a star by 1° every 71.6 years, which can clearly be rounded to 72 for convenience both of transmission through myth and of general calculation).   She does, however, refer to an ancient story explaining that the god Thoth received a prize of $1/72^{nd}$ of a 360-day year for winning a game of draughts with the moon, thus giving Egyptians the five intercalary or epagomenal days added to the 360-day year.   She does deal extensively with precessional numbers, as do de Santillana and von Dechend, and we shall return to them in more detail presently, but the extensive foundations of her argument do not depend upon the presence of such numbers in the most ancient texts.

Sellers points out that the Osiris story is never explicitly laid out in the Pyramid Texts but treated as "a mysterious given" (as she writes on Amazon.com in response to New Age authors citing her as a source for their assertions that precessional numbers appear in ancient texts). Instead, the earliest text we have that systematically lays out the myth of Osiris is from Plutarch of Chaeronea (AD 46 – AD 120). Plutarch was a contemporary of Ptolemy, and lived over a hundred years after the death of Hipparchus, and thus could have learned of precession from sources of which even conventional scholars are aware. It is thus possible (although no evidence supports such a conclusion) that he could have himself inserted the precessional number 72 into his account of the Osiris story.

Of course, as Otto von Neugebauer points out, both Hipparchus and Ptolemy gave 1° every 100 years as their estimate for the lower limit of the precessional constant (never mentioning 72), and such was the impact that Ptolemy's *Almagest* had upon succeeding generations of astronomers that this figure became enshrined as the rate of precession for literally centuries thereafter.

This fact of history argues against Plutarch's insertion of recent astronomical work (even if he was aware of it) into his account of the Osiris myth – to maintain such a theory, we would have to assume that Plutarch or one of his sources knew of these "cutting edge" astronomical observations of Hipparchus and / or Ptolemy, and that they then corrected the work done by those men to a better precessional constant, and then slipped that into the retelling of the Osiris myth (rather than publishing that work in another venue).

Such a theory, while possible, is farfetched. Far more likely is that the Osiris myth, like other myths, was a transmitter for secret astronomical knowledge by the time it came to Plutarch. Whether those secret numbers were already part of the myth as early as the Pyramid Texts, we cannot say for certain. That they were part of the story by the time of Plutarch seems relatively obvious.

However, the important point is that Sellers convincingly demonstrates that the Osiris story contains precessional elements *without* having to resort to the evidence of precessional numbers, since we have no textual evidence of those numbers until Plutarch.

As Plutarch explains the myth in *Of Isis and Osiris, or of the Ancient Religion and Philosophy of Egypt* (or *De Iside et Osiride* in Latin), found in the collection of Plutarch's essays called *Morals* (L. *Moralia*), the Egyptians tell that each of the epagomenal days were the birth-dates of first Osiris, then Arueris ("whom some call Apollo and others the elder Horus," Plutarch says), then Typhon ("who came not into the world either in due time or by the right way, but broke a hole in his mother's side, and leaped out at the wound"), then Isis, and then Nephthys (translations from William Watson Goodwin's 1900 edition of *Plutarch's Morals*, volume 4, page 75). Plutarch uses the name of the Greek monster Typhon for the Egyptian antagonist of Osiris and Horus, Set or Seth.

Plutarch continues with the story of Osiris and Isis:

> And they say that Osiris, when he was king of Egypt, drew them off from a beggarly and bestial way of living, by showing them the use of grain, and by making them laws, and teaching them to honor the Gods; and that afterwards he travelled all the world over, and made it civil, having but little need of arms, for he drew the most to him, alluring them by persuasion and oratory, intermixed with all sorts of poetry and music; whence it is that the Greeks look upon him as the very same with Bacchus. They further add that Typhon, while he was from home, attempted nothing against him; for Isis was very watchful, and guarded him closely from harm. But when he came home, he formed a plot against him, taking seventy-two men for accomplices of his conspiracy, and being also abetted by a certain Queen of Ethiopia, whose name they say was Aso. Having therefore privately taken the measure of Osiris's body, and framed a curious ark, very finely beautified and just of the size of his body, he brought it to a certain banquet. And as all were wonderfully delighted with so rare a sight and admired it greatly, Typhon in a sporting manner promised

that whichsoever of the company should by lying in it find it to be of the size of his body, should have it for a present. And as every one of them was forward to try, and none fitted it, Osiris at last got into it himself, and lay along in it; whereupon they that were there present immediately ran to it, and clapped down the cover upon it, and when they had fastened it down with nails, and soldered it with melted lead, they carried it forth to the river side, and let it swim into the sea at the Tanaitic mouth, which the Egyptians therefore to this day detest, and abominate the very naming of it. These things happened (as they say) upon the seventeenth of the month Athyr, when the sun enters into the Scorpion, and that was upon the eight and twentieth year of the reign of Osiris. But there are some that say that was the time of his life, and not of his reign. 75 – 76.

After this, Plutarch explains that Isis is distraught upon being told of the events by certain children who witnessed the whole thing, and is led to the ark containing Osiris by Anubis, who Plutarch tells us is the son of Osiris by Nephthys, who is wife of Typhon.

Of him [Anubis] she had tidings of the ark, how it had been thrown out by the sea upon the coasts of Byblos, and the flood had gently entangled it in a certain thicket of heath. And this heath had in a very small time run up into a most beautiful and large tree, and had wrought itself [the tree] about it [the chest containing Osiris], clung to it, and quite enclosed it within its trunk. Upon which the king of that place, much admiring at the unusual bigness of the plant, and cropping off the bushy part that encompassed the now invisible chest, made of it a post to support the roof of his house. These things (as they tell us) Isis being informed of by the divine breath of rumor, went herself to Byblos; where when she was come, she sate her down hard by a well, very pensive and full of tears [. . .]. 77 – 78.

At this point, Plutarch relates the activities of Isis in the court of the Queen of the Biblians, where she rubs the queen's women with ambrosia, nurses the queen's son with her fingertips and attempts to singe his mortality away and make him mortal (but is thwarted in this when the queen is too frightened at the sight of her baby on fire), turns into a swallow at night time to fly around the post which holds up the roof, and finally begs for the post and takes it away with her, thus obtaining the chest containing the remains of Osiris. The story continues:

> But when Isis came to her son Horus, who was then at nurse at Buto, and had laid the chest out of the way, Typhon, as he was hunting by moonshine, by chance lighted upon it, and knowing the body again, tore it into fourteen parts, and threw them all about. Which when Isis had heard, she went to look for them on a certain barge made of papyrus, in which she sailed over all the fens. Whence (they tell us) it comes to pass, that such as go in boats made of this rush are never injured by the crocodiles, they having either a fear or else a veneration for it upon the account of the goddess Isis. And this (they say) hath occasioned the report that there are many sepulchres of Osiris in Egypt, because she made a particular funeral for each member as she found them. There are others that tell us it was not so, but that she made several effigies of him and sent them to every city, taking on her as if she had sent them his body; so that the greater number of people might pay divine honors to him, and withal, if it should chance that Typhon should get the better of Horus, and thereupon search for the body of Osiris, many bodies being discoursed of and shown him, he might despair of ever finding the right one. But of all Osiris's members, Isis could never find out his private part, for it had been presently flung into the river Nile, and the lepidotus, sea-bream, and pike eating of it, these were for that reason more scrupulously avoided by the Egyptians than any other fish. But Isis, in lieu of it, made its effigies, and so consecrated the phallus for which the Egyptians to this day observe a festival. 80.

After this, Plutarch relates that Osiris came out of the land of the dead to assist Horus in the battle against Typhon, a battle in which Horus at last prevailed. Typhon being bound was delivered up to Isis, but she "would not put him to death but contrariwise loosed him and let him go," which so angered Horus that he (in Plutarch's words) "laid violent hands upon his mother, and plucked the royal diadem from off her head. But Hermes [as Plutarch calls Thoth] presently stepped in, and clapped a cow's head upon her instead of a helmet" (81). In other versions of the story, Horus actually cuts off the head of Isis, which is replaced by the head of a cow.

This myth was so ancient that it did not need explaining in the Pyramid Texts but was assumed as a known storyline. Already we can see the clear parallels of this ancient myth to the Germanic legends of Amlodhi that Shakespeare drew upon for Hamlet (and thus begin to understand why de Santillana and von Dechend titled their essay *Hamlet's Mill*). We see a son (Horus and Hamlet) avenging the murder of his father (Osiris and the old King Hamlet, each of whom appears to the son after death) by the uncle (the father's brother, Set and Claudius), as well as a certain anger and violence to the mother (Isis and Gertrude).

There is also a Germanic and Icelandic legend concerning Hamlet (or Amlodhi, or Amleth/Amlethus) related by de Santillana and von Dechend in which Hamlet had a mill the stone of which came unhinged and fell into the sea. Recall that the stones of the mill are a helpful metaphor for the spinning action of the heavens (87 – 95).

The account of Isis and Osiris also contains echoes of the Sumerian and Babylonian story of Gilgamesh, in which Isis is presented with the horns of the Bull of Heaven after it is slain (and Isis is often depicted as having cows' horns above a human woman's head in ancient Egyptian and Babylonian statues). It is quite possible, of course, that the Osiris myth came first and the Sumerian and Babylonian accounts are echoes of it.

Plutarch uses the Greek name Typhon for the Egyptian god Set, who (he tells us) is associated with the constellation of the Bear (and the Big Dipper), as Osiris is associated with the constellation of Orion and Isis with the star Sirius (Sothis in Greek and Soped in ancient Egyptian). As we have seen, such myths appear to contain analogies that transmit information about the movements of the heavens, including the phenomena caused by precession. The chopping down of a tree which encloses an ark is an

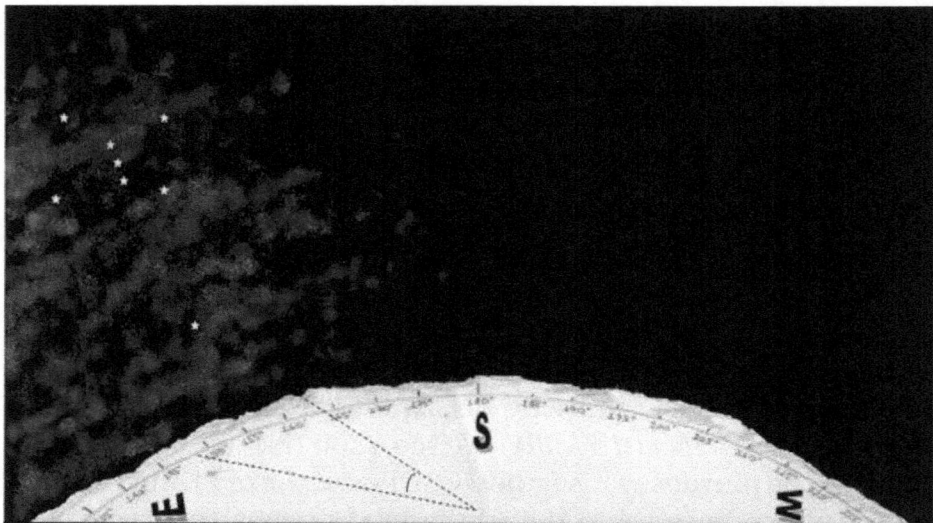

Heliacal rising refers to rising with the sun (the term comes from the name of the sun god Helios). As the earth moves along its path through the year, constellations will rise four minutes earlier than the previous day, because the earth is passing them as it makes its annual circuit. Eventually, constellations near the ecliptic will be rising during the day and will be blotted out by the sun during daylight hours, but as earth continues they will keep rising earlier and earlier until they begin rising just before the sun. Here, Orion and Sirius have risen before the sun and dominate the eastern sky. As the sun rises, the sky will lighten to blue, but they will still be visible until the sky becomes so light that they fade from view. This beautiful heliacal rising of Orion and Sirius now takes place in August. The diagram above depicts the sky on 15 August at 5:45 am. Each morning they will rise four minutes earlier, and thus be 1° further along their way when the sun comes up.

analogy which we find over and over throughout the world. The removal of the pillar that holds up the roof is another aspect of the same image – an analogy that we should be able to understand as a possible reference to the removal of one celestial axis and its replacement with another due to the inexorable action of precession.

Jane Sellers acknowledges that 72 is a precessional number, and includes much evidence for the ancient Egyptian use of precessional numbers in her argument, but her assertion that the Osiris myth analogizes the mechanics of precession does not depend on the inclusion of precessional numbers. In the *Death of Gods in Ancient Egypt*, she explains the story of Osiris's murder and disappearance into the sea in terms of the celestial mechanics behind the motion of the constellation Orion (and the star Sirius).

Orion has an annual cycle of dominating the night sky and then disappearing (during the months in which the sun is between the earth and Orion, such that to see him one would have to look at the sky during the day, when he is invisible due to the sun's power), but this is similar to all other stars far enough away from the celestial north pole to dip below the horizon, as we have seen from our examination of the celestial mechanics in the preceding chapter (remember the analogy of the dining room: if the earth were orbiting a sun in the confines of your dining room, observers on that earth would see different constellations on the "walls" at different points of the orbit, and during some parts of the year the constellations on certain walls would be invisible, as they would be opposite the sun and thus unable to be seen until later in the orbit). The earth's path around the sun means that some constellations will be visible during the night only on some parts of the earth's journey around its orbit, unless those constellations are so close to the north (for dwellers in the northern hemisphere) that they can always be seen (which the Egyptians referred to as the "undying" or "imperishable" stars and constellations, and which correspond to the constellations painted "on the ceiling" of our imaginary orbit inside your living room).

The position of the constellations seen on any particular day each year shift very slightly as the earth progresses around the sun on its annual orbit, moving along by slightly less than one degree per year (one degree being one three-hundred-sixtieth of a circle). Thus, while the constellations move each night from east to west during the night, they also appear one degree further west at the same hour each night. This is because the west is not only the trailing direction in the earth's rotation on its axis, but also from the perspective of earth's progress around the sun on its orbit.

If the earth moves about one degree per day on its path around the sun, thus "passing" the constellations on the "walls" by about a degree each day, then the effect for an observer on earth will be that the earth's turning will reveal that constellation about "one degree sooner" each day. In minutes, this works out to about four minutes earlier each day (since there are 1,440 minutes in a 24-hour day, and since one three-hundred-sixtieth of 1,440 minutes is equal to four minutes). Thus, Orion (or any other constellation) should be about four minutes ahead each day in his progress across the sky, rising four minutes earlier, reaching culmination or the highest point in his passage across the sky four minutes earlier, and then setting four minutes earlier than he did the night

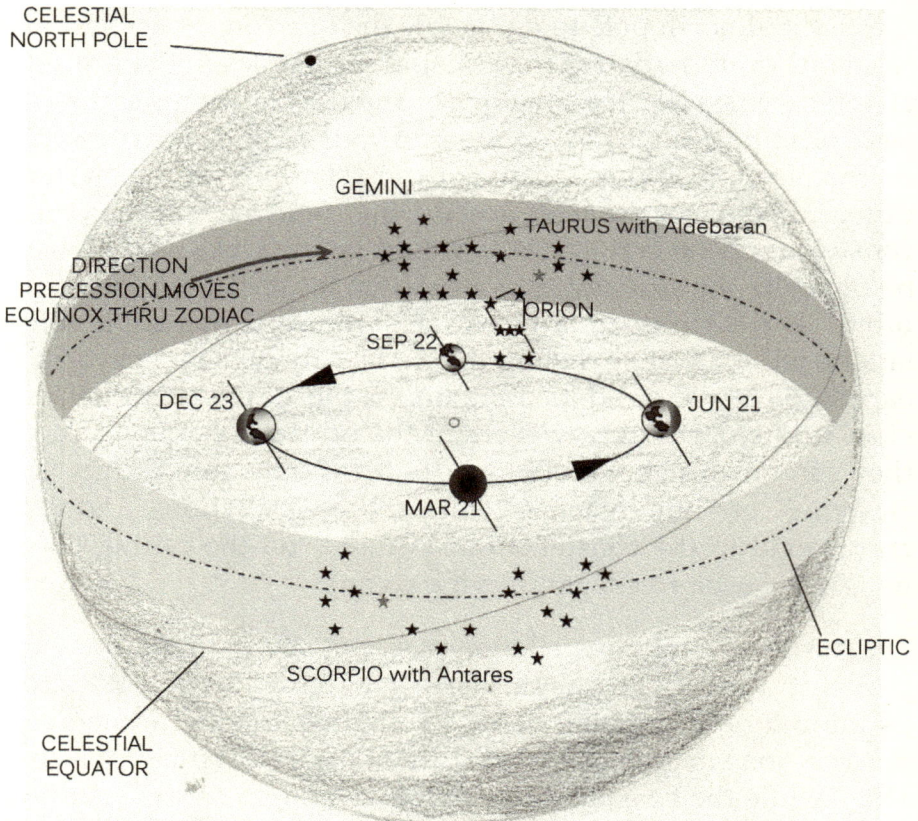

CELESTIAL
NORTH POLE

GEMINI

TAURUS with Aldebaran

DIRECTION
PRECESSION MOVES
EQUINOX THRU ZODIAC

ORION

SEP 22

DEC 23

JUN 21

MAR 21

ECLIPTIC

SCORPIO with Antares

CELESTIAL
EQUATOR

As the motion of precession delays the constellations by 1° every 71.6 years, Orion (not on ecliptic but close to it) will dominate the spring equinox until he is replaced by Taurus. As Orion is "held back" below the equator by precession, it is as if he is being drowned. This celestial drama is encoded in the Egyptian mythology of the murder of Osiris by Set. As Jane Sellers explains, Set is opposed on the opposite side of the celestial globe by Horus the son of Osiris, who was originally associated with the Scorpion and the star Antares.

before. Eventually, this process will mean that on a certain day, he will not rise at all during the nighttime hours. However, as the process continues, his daytime rising will keep getting earlier and earlier, approaching the day when he will beat the sun into the sky by a few minutes. Eventually, there will come a day upon which he is rising just before the sun, far enough before the sun that he (or at least some of his brightest leading stars) can be seen in the sky before the sun's rise blots them out again. (The brightness of a star will determine how far ahead of the sun it must be in order to be visible to an observer on earth -- the brighter the star, the less of a "head start" it needs to get ahead of the sun in order to achieve heliacal rise, and Sirius as the brightest of all the fixed stars needs the least head start of them all).

This day for any star is known as its date of "heliacal rise." It should take place on the same predictable date each year for every star, since on any given day of the year the earth should be in the same spot on its orbit that it occupied on the previous year on the same day. Of course, one could argue that having an imprecise calendar (one that didn't have leap years, for example) would throw this off, but remember that the important day of heliacal rising that is used to name the great "ages" (the Age of Taurus, the Age of Aries, the Age of Pisces and the upcoming Age of Aquarius) was a date determined by the equinox position of the sun, which can be determined using physical stone markers that don't depend on calendars. In a world without precession, the star that would rise just before the sun on the date of the spring equinox would be the same each year, because earth is back at its same point on the orbit.

As any given star continues rising earlier each day, it will eventually be replaced with new stars that rise just before the sun. Obviously, this will cause new stars to occupy the position of heliacal rising every so often, as new ones that were below the horizon when the sun rose "catch up" and overtake the sun and occupy the sky just ahead of it, only to continue along to be replaced by a new set of stars as the year progresses. A different star or constellation would occupy the heliacal rising point on the summer solstice, and on the fall equinox, and on the winter solstice, and these stars would be the same each year, giving way to their successors on the same days as the earth continued around.

This predictable and orderly replacement of heliacal risings should go on in perfect synchronization with an accurate calendar each year. And, within the relatively short span of one human lifetime, this is what actually does appear to happen. However, as we have observed, the wobble of the axis throws a new variable into the process. Even if the earth is at the exact same point on its orbital path, if the axis has wobbled enough, it will alter the cosmic background.

As we saw in chapter 3, this circular motion of the axis is clockwise as we look down on the northern hemisphere, which means that the constellations are delayed in their rising from one year to the next, but only an infinitesimal amount each year. The shift is so slight that even in a lifetime it is barely perceptible – as we have already seen, only one degree every 71.6 years. However, at that rate, after thirty times 71.6 years, a constellation will be rising an entire thirty days later.

Thus, if at some point in the remote past Orion had his heliacal rising before the spring equinox, this delaying action would make him appear to be lower and lower over the years on the same day, until at some point he was delayed enough that the preceding constellation would be in the position of heliacal rise on the spring equinox. The fact that the action of precession delays a constellation and causes the *preceding* constellation to occupy the position instead leads to the phenomenon's name "precession."

This has happened to Orion and Sirius (Sothis) since the days of pre-dynastic Egypt (when Sellers believes that the Osiris legend was first created). Orion's heliacal rising has slipped all the way from the spring equinox to August between their day (millennia ago) and ours. He has been delayed -- "held down," so to speak, beneath the horizon. In other myths, this action of holding beneath the horizon will be compared to drowning. The same has happened to the constellations dominating the equinoctial and solstitial dates as well – from the Age of Gemini, to the Age of Taurus, to the Age of Aries, to the Age of Pisces (now coming to an end itself).

The process depicted in the diagram above from the perspective of an earth observer (in the northern hemisphere at the latitude of ancient Egypt – or central California). This diagram is now familiar from our extensive examination of these phenomena, but for simplicity depicts only the principle constellations which Sellers will treat in her theory for the origins of the Egyptian mysteries: Orion (along with Sirius, not shown), Taurus, and Scorpio.

As the diagram illustrates, the shift in the sky eventually caused Orion and Sirius to fail to make their customary heliacal rise on the date of the spring equinox, during the time of shifting into the Age of Taurus. When Taurus (or simply its brightest star, the reddish Aldebaran) replaced Orion, this was encoded in myth as Set or Seth (whom Plutarch names Typhon) slaying his brother Osiris.

As Jane B. Sellers explains:

> I am postulating the creation of specific myths to deal with distressing alterations in the sky, followed by an artificial duality, or symmetry, imposed, not just on the deities, but on geographical centers of worship, and this duality remained a constant in Egyptian affairs throughout its history. It was a harking back to a wonderful Golden Age, now lost; an age when the skies had had a magnificent balance. **Taurus**, with red Aldebaran, marked the **spring** date of equal balance, and **Scorpius**, with bright Antares, marked the **autumn** date. 88 [bold emphasis in the original].

As part of the extensive textual support Sellers summons to advance this thesis, she cites a passage from the Shabaka stone (engraved in the reign of King Shabaka of 780 BC, who said he copied it from a far more ancient source, which scholars had once thought to have been as old as 2664 BC or even 3110 BC, although current scholars have cast some doubt upon these claims of Shabaka, and so we can only say that it is at least as old as 780 BC or so).

In the Shabaka text, which is commonly referred to as "The Memphite Theology," Sellers finds these relevant details: The rivals Seth and Horus have battled for eighty years, and the sky deity Geb has called together the Ennead, or the Nine gods, for a trial to decide between them. As Sellers points out, it is suspicious that Seth should be considered at all for the position of successor to Osiris, since everyone knows he is the murderer. Suspicious, that is, unless the myth has an astronomical origin. Perhaps because the star associated with Seth (Aldebaran) and his constellation (Taurus) succeeded Orion in the heliacal rising, he appears to have a close claim on the succession, and if this is the reasoning, then it is a strong piece of evidence in support of a stellar and precessional background for the original Osiris – Set – Horus mythology. Geb begins the trial:

> Geb, Prince of the gods, commanded the Nine gods gather to him. He judged between Horus and Seth: he ended their quarrel. He placed Seth as king of Upper [**Southern**] Egypt, up to the place in which he was born, which is Su. And Geb placed Horus, king of Lower Egypt In the Land of **Northern** Egypt, up to the place in which his father was drowned, which is Division-of-the-Two-Lands. Thus Horus stood over one region, and Seth stood over one region. They made peace over Two Lands. That was the boundary of the Two Lands. Sellers, 86.

To be clear: these "two lands" were not primarily terrestrial regions, but portions of the sky. Horus rules the end of the Milky Way dominated by Scorpio and Antares, as well as the section of the sky "north" of the Milky Way – on the same side of the Milky Way as the celestial north pole. This celestial region is given to Horus because his star, Antares, is on the same side of the Milky Way as the celestial north pole. Seth receives the end of the Milky Way dominated by Taurus and Aldebaran, as well as the section of the sky "south" of the Milky Way, because Aldebaran is on the southern side of the Milky Way, the section that does not contain the celestial north pole.

Also, these "two lands" designate portions of the year. As de Santillana and von Dechend have convincingly argued, the well-known concept of "the four corners of the earth" still found in fairy tales and folktales does not necessarily derive from a time when people believed the earth was a flat square – even if you thought the earth was flat, it would make more sense to picture it as a flat disc, which is what it resembles from our perspective. In fact, these authors argue, the ancient concept of an "earth" was more expansive than what we think of as "earth," as it included both what we call "earth" as well as the heavens. Giorgio de Santillana and Hertha von Dechend explain:

> On the zodiacal band, there are four essential points which dominate the four seasons of the year. They are, in fact, in church liturgy the quatuor tempora marked with special abstinences. They correspond to the two solstices and the two equinoxes. The solstice is the 'turning back' of the sun at the lowest point of winter and at the highest point of summer. The two equinoxes, vernal and autumnal, are those that cut the year in half, with an equal balance of night and day, for they are the two intersections of the equator with the ecliptic. Those four points together made up the four pillars, or corners, of what was called the 'quadrangular earth.'

> This is an essential feature that needs more attention. We have said above that 'earth,' in the most general sense, meant the ideal plane laid through the ecliptic; meanwhile we are prepared to improve the definition: 'earth' is the ideal plane going through the four points of the year, the equinoxes and the solstices. Since the four constellations rising heliacally at the two equinoxes and the two solstices determine and define an 'earth,' it is *termed* quadrangular (and by no means 'believed' to be quadrangular by 'primitive' Chinese, and so on). 62.

Thus, by allotting Seth and Horus their respective posts, which correspond to certain sections of the celestial sky, Geb has also given them portions of the year, or portions of the "quadrangular" earth: Horus has received the autumnal equinox, and Seth has received the vernal. Between them, the year could be said to have been in balance, but it was more of the balance of a tug-of-war: they were striving between one another for dominance until the judgment of Geb and the Ennead.

As Jane Sellers relates, the balance between Seth and Horus which obtained after the death of Osiris – Seth ruling the "south" and the

Spring Equinox, and Horus ruling the "north" and the Autumn Equinox – did not last. Precession put an end to this balanced situation, which, even in its conflict, had an almost miraculous symmetry (particularly in the twin red stars, as Sellers points out in her earlier quotation).

In the description of the judgment of Geb and the Nine gods, the Memphite Theology relates:

> Then it seemed wrong to Geb that the portion of Horus was like the portion of Seth so Geb gave Horus his inheritance, for he is the son of the firstborn son.

> Geb's words to the Nine Gods: I have appointed Horus, the firstborn. Him alone, Horus, the inheritance. The inheritance belongs to the son of my son, Horus, the Jackal of Southern Egypt . . . the first born Horus, the Opener of the Ways. The son who was born Horus, on the Birthday of the Opener of the Ways.

> Then Horus stood over the land. He is the uniter of this land, proclaimed in the great name: Ptah, South-of-his-wall. Lord of Eternity.

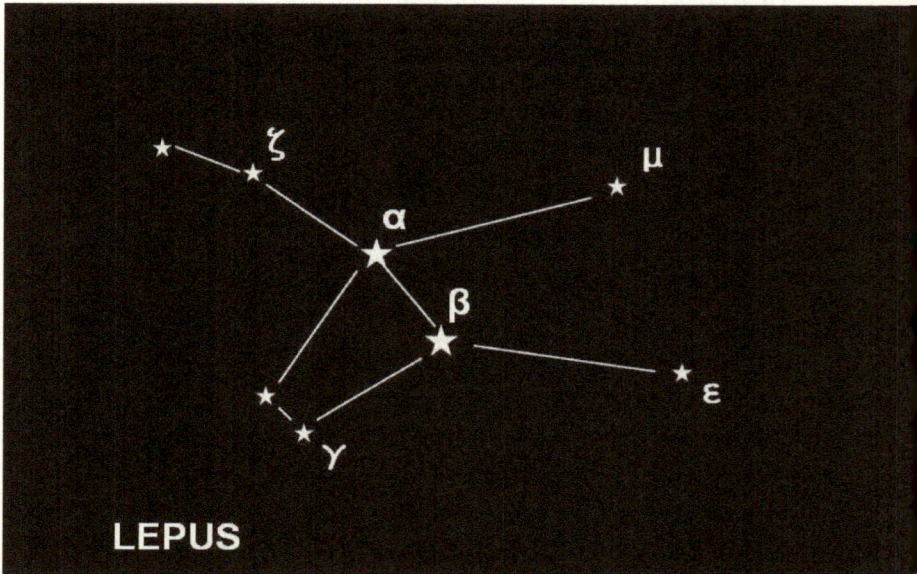

LEPUS

Then sprouted the two Great Magicians upon his head. He is Horus who arose as King of Upper and Lower Egypt, who united the Two Lands in the Name of the Wall, the place in which the Two Lands were united. Reed and papyrus were placed on the doubledoor of the House of Ptah. That means Horus and Seth, pacified and united. They fraternized so as to cease quarreling in whatever place they might be, being united in the House of Ptah, the 'Balance of the Two Lands,' in which Upper and Lower Egypt had been weighed. Quoted in Sellers, 86 – 87.

Sellers explains that Horus is given the inheritance, but not as a new heliacal rising position, but rather a position in "the bark of the sun," a shift from a stellar deity to a solar one, and this in part explains the seemingly puzzling existence of "two" Horus deities in ancient Egyptian religion, an old one and a new Horus-son-of-Isis, identified with the sun (Sellers 90).

Thus, there has been a "big switch" – Horus has taken over the entire "inheritance," and Seth has been dispossessed. Horus is now the King of Upper and Lower Egypt. Seth's dispossession

http://commons.wikimedia.org/wiki/File:Stela_of_Tuthmosis_I.jpg

is further emphasized by the legends that Horus seizes Seth's testicles, which metaphorically represents a seizing of someone's inheritance, since in the human realm our heirs and our estate are typically our offspring. In some Egyptian texts, there are indications that Seth has now been sent all the way to the other side of the heavens, to the side of the Scorpion, since he is in one text given the scorpion goddess Selket as his consort, who becomes with him the mother of a brood of serpents (Sellers 106).

Jane Sellers provides remarkable layers of evidence to support her thesis, including solar eclipse imagery found in the most ancient texts which is of critical importance to the concept of the "Eye of Horus" found in Egyptian funerary texts. She provides ancient Pyramid Texts which support her theory that a complete solar eclipse was analogized as Seth obscuring and defiling the face of Horus, but that the dramatic "diamond ring" sunbursts which appear just prior to "totality" and just at the end of it – the Eyes of Horus – represent the triumph of Horus (the solar deity) over Seth.

She also notes that in the text known as the *Contendings of Horus and Seth* (preserved in a papyrus from the reign of Ramses V, generally believed to have ruled from 1149 BC – 1145 BC), the now-deposed Seth is given a new position in the sky beneath Orion, commonly associated with the stars of the constellation known today as Lepus (the Hare).

The appearance of the constellation Lepus is most remarkable when considered in context with certain details of Seth's appearance. The constellation (depicted above) contains eight stars and is located almost directly below the feet of the constellation of Orion. If, as I believe in agreement with Jane Sellers, Giorgio de Santillana and Hertha von Dechend, the Osiris mythology refers to celestial figures and events, then the declaration that Seth must be placed "under Osiris" would associate him with these stars, and Sellers asserts that "This visible location, in all probability, was the most popular with the people" (107).

The shape of the stars below Orion alone seem to support this reading of the mythology. Seth has very distinctive features, often described as a combination between a dog and an ass. Because (unlike most other Egyptian deities) his head does not seem to be an exact fit with any animal, it is often called the head of a "Seth beast" (de Santillana and von Dechend 163).

The resemblance in the constellation of Lepus to the oft-occurring depictions of Seth is fairly obvious, as the constellation itself appears to have long ears and an elongated snout:

Compare this with the depictions of Seth in this stele from the reign of Thuthmoses I (believed to have ruled from 1506 BC – 1493 BC), shown above on the page opposite the constellation Lepus.

Here we can clearly see the long ears and narrow snout that are remarkably similar to those found in the constellation below Orion (the constellation "carrying Osiris").

Plutarch's descriptions of the Egyptian traditions surrounding Set aka Typhon are very revealing and appear to provide additional support for Sellers' theory, although she does not cite this portion of Plutarch in her argument.

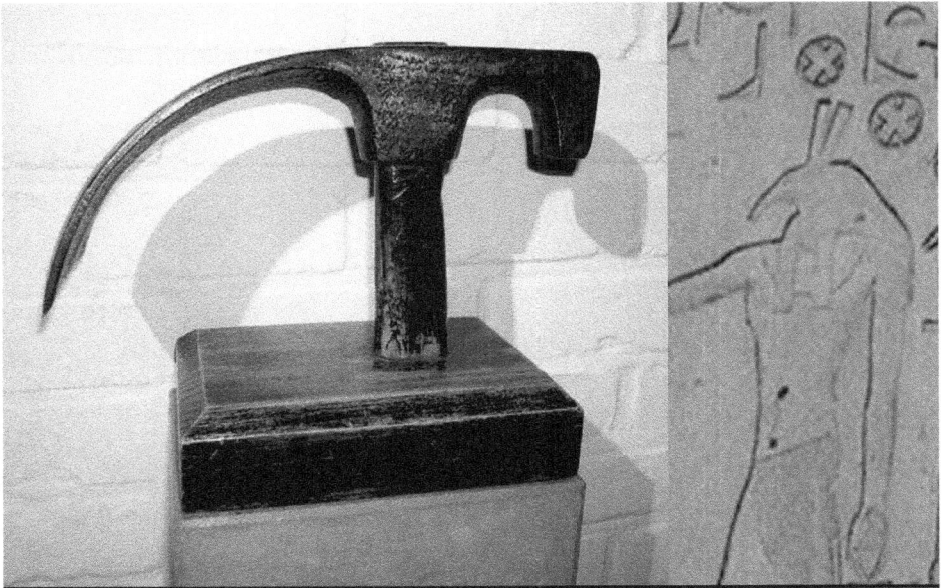

Detail from the previous depiction of Seth juxtaposed with the head of an adze, clearly showing the resemblance to the distinctive features of the "Seth-beast." The adze was not only known to the ancient Egyptians but featured prominently in their religious symbology, especially in the ceremony known as the "opening of the mouth," during which the mouth of the mummy was touched by symbolic foods and implements, including the adze.

Plutarch relates:

> Osiris therefore and Isis passed from the number of good
> Daemons into that of Gods; but the power of Typhon being
> much obscured and weakened, and himself besides in great
> dejection of mind and in agony, as it were, at the last gasp,
> they therefore one while use certain sacrifices to comfort
> and appease his mind, and another while again have certain
> solemnities wherein they abase and affront him, both by
> mishandling and abusing such men as they find to have red
> hair, and by breaking the neck of an ass down a precipice (as
> do the Coptites), because Typhon was red-haired and of the
> ass's complexion. Moreover, those of Busiris and Lycopolis
> never make any use of trumpets, because they give a sound
> like that of asses. And they altogether esteem the ass as an
> animal not clean but daemonic, because of its resemblance to
> Typhon; and when they make cakes at their sacrifices upon
> the months of Payni and Phaophi, they impress upon them
> an ass bound. William Watson Goodwin translation, 90-91.

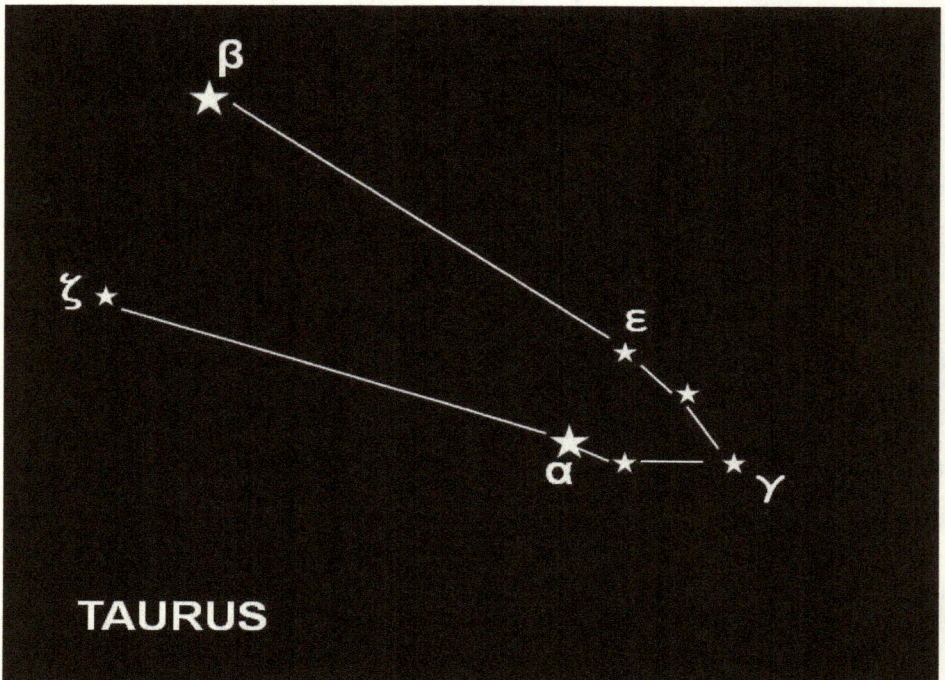

TAURUS

Here we have an ancient source associating Seth (Typhon) with the ass (an animal with long ears and a narrow snout, and one that the reader is invited to bear in mind while considering the outline of the stars of the constellation Lepus depicted above), as well as with the color red (a color associated both with the color of the animal with which Plutarch says Seth is associated, but also the distinctive color of the star Aldebaran, an association central to the thesis advanced by Jane B. Sellers).

Another remarkable resemblance to the features of Seth and the Seth beast is found in the shape of the implement known as the adze. A heavy tool with a curved blade and a long handle intended for use with both hands, an adze is generally used to smooth heavy pieces of timber.

Jane Sellers has noted that the sharp "V" angle of the brightest stars in Taurus – with the important red star Aldebaran, or alpha Tauri, at the center of the "V" – resembles the adze, as does the front end of the constellation of the Scorpion (95). Her thesis posits that Seth was associated with Aldebaran and this "V"-shaped set of stars, which took over the spring equinox after the "drowning" of Osiris (the failure of Orion to rise above the horizon at the appointed time). Thus, the fact that the features of Seth bear a powerful resemblance to the shape of an adze may be very significant (see the diagrams on preceding pages).

The resemblance is even more striking in the statue of the coronation of Ramses III, which can be seen in the Egyptian Museum in Cairo (the Museum of Egyptian Antiquities). There, Horus and Seth stand beside the king, each with a hand upon his crown. Seth is to the king's left, and Horus to his right. Ramses III is generally believed to have reigned from 1186 BC – 1155 BC.

The adze was by no means unknown to the ancient Egyptians, but in fact featured prominently in their religious rites, in particular the ceremony known as the "opening of the mouth" (or the "mouth-opening ceremony").

This ceremony was of utmost importance and there exists an entire body of ancient texts known as "The Book of the Opening of the Mouth." These have been described extensively by many scholars, including Sir Ernest Alfred Thompson Wallis Budge (1857 – 1934).

In his 1909 study, the *Liturgy of Funerary Offerings: The Egyptian Texts with English Translations*, E. A. Wallis Budge describes the importance of this ceremony:

> In the case of the "Book of Opening the Mouth," the object of the recital was, in the earliest times at least, to bring about the reconstitution and resurrection of the dead man, and even in later times, when the work was recited before a statue, on which the accompanying ceremonies were performed, the idea of the Egyptians on this matter remained unchanged. It must be remembered also that the Egyptians intended by means of ceremonies and formulae to bring back the Ka, or double, either to the dead man, from whom it had been temporarily separated, or to a statue which represented him; and when this had been done they believed it to be their bounden duty to provide meat and drink for its maintenance. It was the Ka and the heart-soul (Ba), not the spirit-soul (Khu), which fed upon the offerings, and if meat and drink of a suitable character, and in sufficient quantity, were not provided for them, these suffered from hunger and thirst, and if the supply of offerings failed, they perished by starvation. The texts make it quite clear that the Egyptians believed in a dual-soul; one member could not die, but the other only lived as long as it was fed with offerings by the living and provided with an abode, i.e. a statue. Offerings were brought to the funerary chapels and tombs daily, and additional gifts were presented on the days of all great festivals. viii – ix.

The adze features prominently in this important ceremony, as does the "foreleg of the ox" or "foreleg of the bull," which is pictorially depicted and described often in the texts describing the opening of the mouth. As Jane B. Sellers explains in a footnote:

> In certain illustrations, the adze is replaced with a Foreleg of an Ox, in others, both are represented on a table that holds the ceremonial tools. The identification with their constellation, 'Meskhetui,' believed to be the same as the Great Adze or the 'Foreleg,' or 'Thigh of Seth,' and our Big Dipper, seems firmly established. 328.

Scene from the side of throne found in the tomb of Sesostris I depicting Horus (left) and Seth (right) turning the celestial drill or mill. As we have seen, there is extensive evidence that the struggle between Seth and Horus is a mythological embodiment of celestial events. The turning of the celestial axis would move the colures and shift the equinoxes and solstices to new constellations along the ecliptic.

The association of the thigh of the ox with the thigh of Seth reinforces Sellers' argument that Seth and the constellation of the Bull (and especially the V-shaped stars with Aldebaran at their crux) played a central role in the origins of the Egyptian religion.

We find further possible support for this association between the adze and Seth in the shape of the *was* scepter carried by Seth and resembling at its top his features. It also resembles an adze, and the conflation of Seth with the head of an adze-like instrument appears to reinforce the connection between the two.

Further support for Sellers' thesis is found in the depiction of Seth and Horus in scenes commonly interpreted (de Santillana and von Dechend say "continuously mislabeled") by the conventional theorists and Egyptologists as "the uniting of the two countries." We have argued that the "uniting" is not primarily to be understood as a mythologizing of any historic "conquest" or unification of two terrestrial pieces of real estate (although contemporary academics, seeking to find political, economic, and socio-cultural "power" relationships in every aspect of culture, immediately tend towards such interpretations) but rather a depiction of celestial concepts.

The diagram on the preceding page clearly supports such an interpretation. It is from a throne belonging to the reign of Sesostris I (also known as Senusret I, reigning from 1971 BC – 1926 BC).

In it, we see that the contending of Seth and Horus involves the celestial axis-pole, envisioned as a drill or churning-stick (an image found again and again in world mythologies) and turned by means of a rope or cord between them. While the element of uniting Egypt is present (the scenes are, after all, found on thrones and coronation ceremonies, indicating the earthly as well as heavenly power of the pharaoh), the presence of Seth and Horus points to a precessional symbolism. Their arrangement around a central axis indicates an understanding similar to that encoded in the symbols of Sumer, Babylon and the Omaha tribe that the turning (or the felling, in the case of the miraculous cedar) of the axis turns the equinoxes and drives precession.

While this throne scene comes from the twentieth century BC, even more ancient depictions of Seth and Horus on one side and the other of a king (as in the coronation of Ramses III depicted in the well-preserved statue in the Egyptian Museum) are found in the temple of the necropolis at Saqqara of the earliest tomb containing Pyramid Texts, the tomb of Unas or Unis (believed to have ruled from 2353 BC – 2323 BC).

In the fragmentary remains of the scene, Horus and Seth are standing on either side of King Unas, each with one hand upon his shoulder and the other placing the double crown upon the king's head. This double crown is usually seen as representing

The similarities to the depiction of Seth and Horus pulling reeds tied to a central "drill" post (the celestial axis) in the illustration below and numerous other examples of artwork depicting scenes from the ancient Indian Puranas of the "churning of the ocean of milk" are striking. Note the central axis (in the Hindu texts, Mount Mandara) again has a base narrower than its top, just as we saw in the detail of Horus and Seth in the throne of Sesostris I. Note the "Typhonian" (Seth-like, elongated) features on the creatures on the right side side in the tug-of-war (known as the Asuras in the Hindu texts, struggling against the Devas on the opposite side of the axis). These are arrayed on the same side that Seth occupied in the Sesostris throne-scene (the left-hand side, if considered from the perspective of the axis itself, and in this case the left-hand side of the avatar of Vishnu seated on top of the pillar). Could Vishnu's four arms have anything to do with the concept of the "quadrangular earth" discussed earlier in this chapter?
Image: Wikimedia commons.

his authority over both Upper and Lower Egypt, which is certainly part of the symbolism, but probably also represents celestial authority, as the presence of the two gods who stand over the two equinoxes seems to indicate.

In the Pyramid Texts themselves found within the tomb of Unas, there are passages which reinforce this interpretation and the

Another depiction of the churning of the celestial ocean, again with clear parallels to the depiction of Seth and Horus in the same act. Again the bestial features of the beings on the right indicate their connection with the lineage of Seth. In this scene, the central axis rests upon the tortoise avatar of Vishnu, who also sits atop the axis with his legs crossed. Celestial imagery such as the moon and the bow indicate that this scene involves the motion of the heavens.
image: Wikimedia commons.

One more Old World example of the same celestial churning, this time from Cambodia and worked in sandstone. While the two Indian examples previous to this were from the 19th century AD (although depicting events from ancient Puranas), this dates to the 12th century AD. Again, we note the base of the central pillar is narrower than the top, and set on the back of a tortoise. Vishnu this time is wrapped around the pole itself, although he still has four arms. The beings on the right as we face it are demons, and those opposite are gods.

celestial imagery. At one point, for instance, the hieroglyphs declare: "For Unis has tied together the peppergrass cords, Unis has united the skies" (Allen and Der Manuelian 61).

Again, this concept of the unification of the sky appears to be strong support for the identification of the imagery of Seth and Horus with the majestic rotation of the sky, and (along with the depictions of Seth and Horus above) introduces the idea that their conflict creates a sort of balance, that the opposition of their forces turns the celestial drill.

The ancient Egyptian religion is a deep pool into which we could continue to descend much further without ever reaching the bottom. The convincing arguments in Sellers' thesis concerning the connection of the Eye or Eyes of Horus with the events of a total solar eclipse unlock many fascinating aspects of the Pyramid Texts and other ancient funerary texts from Egypt, but bear less directly on the aspects of precession we are discussing.

It seems clear that extensive evidence supports the thesis articulated by Jane Sellers (and intimated with less specific detail in *Hamlet's Mill*) that the myth of the death of Osiris and the contending of Seth and Horus has its origins in cosmic events and reveals clear awareness of precession in extreme antiquity – certainly prior to 2323 BC when Unas died and the first Pyramid Texts were inscribed, and possibly as early as the period from 6700 BC to 6900 BC, when the failure of Orion and Sirius to rise at the vernal equinox would have taken place.

To sum up, it seems that the myth of Set, Horus and Osiris encodes precession: the slaying of Osiris by Set encodes the delaying of the customary heliacal rise of Orion and the rise of the adze-shaped stars of Taurus in his place. The connection to the chopping down of the world-axis is encoded in the tree which grows around the casket of Osiris, which is later chopped down and turned into a pillar in a palace, and then given to Isis at her request (we shall explore the importance of this pillar in the next chapter as well).

Jane B. Sellers does not necessarily argue for the existence of an advanced ancient civilization. It *is* possible to argue that the early pre-dynastic dwellers in the land of Egypt observed stunning total eclipses and noted the eventual failure to rise of Sirius and Orion on the appointed day each year, and without even a great amount of sophistication concocted legends to explain and transmit the amazing celestial events, perhaps even without great understanding of the causes of those events.

The thesis of Jane B. Sellers does not actually require an ancient and advanced civilization. Perhaps, as she suggests, the legend that Osiris was the first to teach men to cultivate food crops indicates that the spring heliacal rising of his constellation Orion signaled the time to plant to those earliest converts from primitive hunter-gathering to settled agriculture.

And yet, there are hints – which Jane Sellers acknowledges, while declaring that she is "hardly a 'New Ager'" and is not exactly keen to see her work being taken as support by those of a "New Age" bent – that the complex evidence in the mythology suggests an ancient sophistication beyond what anyone would expect from such an early source, even while we acknowledge that they had brains like ours and the ability to ponder the heavens as well as anyone else.

For one thing, there is the precision involved in calculating precession to the point of knowing the importance of the number 72, which even Hipparchus would not have recognized as a

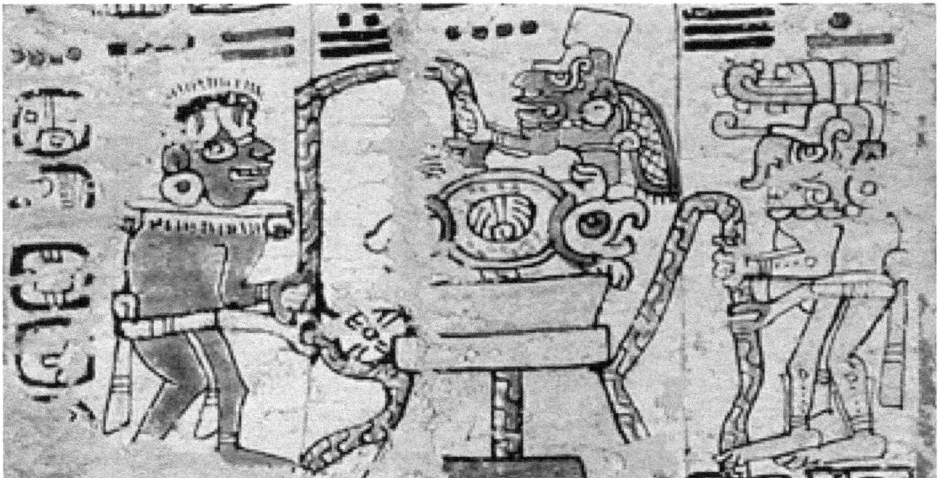

Detail from the Codex Tro-Cortesianus. The similarities to the scene of Seth and Horus are too numerous to be coincidental.

precessional number (we will examine the precessional numbers presently).

As Sellers says in her text, even if we concede that Plutarch was the first to actually record the fact that the companions of Seth who nailed Osiris into his casket numbered 72, the question remains (in her words, and in all capital letters with her italics for emphasis) "Since Nicholas Copernicus, who lived from 1473 to 1543 AD, has always been credited with giving the correct numbers, (although Arabic astronomer Nasir al-Din Tusi, born 1201 AD, is known to have fixed the precession at 51), we may correctly ask, with justifiable astonishment: *JUST WHOSE INFORMATION WAS PLUTARCH TRANSMITTING?*" (196). Known astronomers prior to Plutarch who had determined the precessional constant to be 72 are nonexistent.

But that is not all. Perhaps even more difficult to explain is the fact that the elements of the Osiris-Isis-Seth-Horus series surface in mythology around the world, and not just in cultures that the conventional framework of history could conceive of as having some sort of cultural contact with ancient Egypt (India, Sumer, and Babylon, for instance).

Giorgio de Santillana and Hertha von Dechend point out the obvious similarities between the depiction of Horus and Seth reproduced above and the scene described in ancient Hindu Indian literature and depicted in the scenes of the "churning of the sea of milk," shown on pages 118 - 119.

While the first scene above does not clearly include the tortoise avatar who has volunteered to support the celestial axis, he may perhaps be lurking under the water at the base of the axis. The second and third Hindu scenes do depict him, as do both of the similar scenes depicted in Hamlet's Mill on pages 164 and 165

It seems obvious that these scenes share a common source with the artwork depicting Seth and Horus, or that there was cultural contact and sharing between ancient Egypt and ancient India. The suggestion that these ways of expressing and depicting the phenomena of the heavens arose independently – that they are somehow obvious and would suggest themselves to cultures with no contact with one another – is ludicrous when the numerous common elements are considered.

How then does the conventional theory explain the presence in a Mayan codex of a scene with many of the identical details depicted in the scenes we have just examined? In the longest Mayan codex remaining to us (and there are only four, most of the others having been destroyed by the Spaniards in the 1600s), a codex known as either the Madrid Codex or the Codex Tro-Cortesianus, the image on "page 19" of the codex contains the familiar serpent rope found in the Hindu adaptations of the Seth-Horus scene, as well as the turtle or tortoise, and the figure on the right of the scene has distinctly "Typhonian" features (elongated, hooked nose, monstrous aspect). There is even a "hot-cross bun" (the *kin* symbol of the sun), which we saw featured prominently in the ancient Egyptian depictions of Seth in both the stele from Tuthmoses I and the throne from Sesostris I.

The presence of this scene in a Mayan codex presents insurmountable difficulties for the conventional theory of man's progress and history.

Such conventional theories do not make room for a culture capable of crossing the oceans and disseminating its religio-astronomical imagery to dwellers in the Americas. As we noted at the outset, this possibility has been argued extensively by many whom the academic "establishment" would like to ignore (in the same way that Scotland Yard is always wishing they could ignore Sherlock Holmes in the stories of Arthur Conan Doyle). Graham Hancock has been one of the most articulate and prolific of these analysts, whom the keepers of academic probity would like to label a "fringe" element.

Hancock has posited a technically-advanced civilization of unbelievable antiquity, unknown to conventional academia, and destroyed perhaps by a catastrophe in the ancient past, possibly having to do with the catastrophic melting of the great ice sheets that blanketed much of the largest northern landmasses during the last great Ice Age.

However, the geological evidence discussed by Walt Brown in his hydroplate theory may provide a much more compelling fit with the evidence we see in mythology and archaeology which argues for an ancient culture or cultures that had a very sophisticated knowledge of precession (a knowledge that conventional academia refuses to countenance, as it would upset their cherished theory of man's slow progress from hunter-gatherer to modern space age) as well as the ability to traverse the great oceans long before later civilizations lost these skills.

The evidence we have examined in this chapter not only reinforces the evidence that Dr. Brown finds in the geological record, but also provides glimpses of an alternative history of mankind's past which accords well with the description of events found in the hydroplate theory of Walt Brown. We shall see in future chapters how the clear evidence of ancient knowledge of precession may also be explained by the events in the hydroplate theory as well.

The Djed Column, the World-Tree, and the Great Jars Under the Earth

If the foregoing connections between the demonstrably celestial battles of Seth and Horus and the far-flung manifestations of the identical thematic elements in Hindu and even Maya cosmology do not convince the skeptical reader, we shall visit several more from elsewhere on the globe.

Remember that in the Egyptian Osiris myth, the casket of the slain god was enclosed in a magnificent tree (at one point Plutarch identifies it as an erica tree, often referred to as a heath or heather tree in English) which was admired by the king of Byblos and made into a pillar in his palace. The faithful Isis, tracking her beloved there, would turn into the form of a swallow at night and fly around the post. She eventually recovered the post, an event of great significance. This singular pillar was called the *Tet* (or, as it is more commonly rendered by modern scholars, the *Djed*) and it clearly held great importance in the Egyptian religion. It was depicted with certain distinctive features (see diagrams below), some of which manifest themselves in the central pillar or pole-feature in the Mayan Codex and Hindu Amritamanthana scenes we examined in the previous chapter.

Jane B. Sellers notes that utterance 574 of the Pyramid Texts is addressed to the Sacred Tree, which in the text is specifically identified with the Djed pillar:

> Hail to you, you tree which encloses the god,
> under which the gods of the Lower Sky stand,
> the end of which is cooked, the inside of
> which is burnt, which sends out the pains of

death: may you gather together those who are
in the Abyss, may you assemble those who are
in celestial expanses. Your top is beside
you for Osiris when the djed pillar of the
Great One is loosed.

Sellers 138. Citing Faulkner, *Ancient Egyptian Pyramid Texts*, 229.

Sellers convincingly argues that this tree, guarded by a female goddess and associated with the death of her lover, has numerous echoes throughout the ancient mythologies of other cultures.

For instance, she points to the work of James Frazer (1854 – 1941), author of the *Golden Bough,* who identified the connection with the sacred tree in the grove of Diana of the Wood in Roman mythology. Frazer was fascinated by the recurrence of myths with a goddess, a wood, and a god or hero who dies a violent death in the context of the goddess and the wood. Frazer notes that with Diana in the

The King of Byblos gives the Djed to Isis (left); from a bas-relief in the tomb of Seti I (1290 or 1294 BC - 1279 BC). The Djed scene with Isis and Nephthys (right) from the Papyrus of Ani (c. 1240 BC).

grove in the Roman legend was the god Virbius, who was associated with Greek Hippolytus, a hero dragged downward to his death when his horses were frightened by a bull sent by the sea god.

The possible parallels to the Osiris myth are fascinating. Sellers' thesis is that the Osiris myth encodes the celestial "death" of the constellation Orion as the ruler of the spring equinox. The murderer, Set or Seth, is associated with the star Aldebaran and the adze-like shape of the constellation Taurus, the bull. Hence, in the myth of Diana and Virbius, we have a god whose demise is initiated by a bull, and a goddess who guards a sacred tree.

The association is that the death of the god is somehow connected to the sacred tree – because it is, of course, the shifting of the central celestial axis which causes the precession that "drags" the god down to his "death" (prevents his rising at the proper time to rule the crucial equinox).

Once this connection is unlocked, the parallels to other gods who were the consorts of powerful goddesses and who met with a violent end come crashing in on top of one another.

There is the Babylonian god Tammuz, the lover of Ishtar, who was cruelly slain and his bones ground in a mill. We have already seen the image of the turning heavens as a great upper millstone, and seen that the hub of this millstone is apparently out of joint, causing it to "jump its track" and destroy the successive ages. Just like Isis, Ishtar pursues her slain lover even in death. Sellers records the lament for Tammuz found in Frazer's *Golden Bough*:

> At his vanishing away she lifts up a lament,
>
> Oh my child! At his vanishing away she lifts up a lament,
>
> My Damu! At his vanishing away she lifts up a lament,
>
> My enchanter and priest! At his vanishing away she lifts up
>
> a lament,
>
> At the shining cedar, rooted in a spacious place,
>
> In Eanna, above and below, she lifts up a lament.

Sellers 140, citing Frazer 380.

The connection to the shifting celestial axis is revealed here by the presence of the cedar, which we earlier argued represented the celestial pole in the Gilgamesh series. The death of the lamented god, pursued by the goddess, is seen to be connected in some way with this sacred shining cedar – because it is caused by precession.

Again, the myth of Venus and her young lover Adonis appears to fit the same pattern. Adonis resists the advances of Venus in a grove and is slain by a raging boar. Frazer notes that it was in the shape of a black pig that Typhon injured the eye of Horus, and that it was while boar-hunting that Typhon came across the casket of Osiris that Isis had hidden before he pulled out the body and cut it into fourteen pieces. It is also noteworthy that the boar which kills Adonis is sent by jealous Ares (who always pursues Aphrodite but is often spurned by her), and in some versions of the legend that the boar is in fact Ares himself in disguise. Note that the red star in the Scorpion (associated with Horus) is known as Antares – the "anti-Ares" or rival to Ares, because of that star's bright red color. If Horus is the rival of Ares, and if it is Ares or his boar that slays Adonis, then Adonis the beloved of Venus can be seen as analogous to Osiris the beloved of Isis, and Ares in this myth (or his boar) analogous to Seth. It is also worth noting that Adonis spurns the advances of the goddess, a pattern we first noted in the Gilgamesh legend in which the hero spurns Ishtar or Inanna.

Frazer argues that Adonis and Tammuz are one and the same – the Greeks merely converted the Semitic title *Adon* or "lord" (a title which was applied to Tammuz) into a proper name and gave it to the slain god in their version of the ancient story.

And, Frazer notes, there was a myth in which Aphrodite pursued Adonis into Hades to ransom her lover from Persephone. Aphrodite had hidden Adonis (in a chest, in a clear parallel to the Osiris story), and given the chest to Persephone, the consort of Hades. When Persephone opened the chest, she wanted to keep Adonis for herself.

Connections between the myths of Egypt and the myths of

ancient Greece may not, after all, seem so surprising, and the conventional framework of history has little trouble with them, simply asserting that the later Greeks were influenced by their near neighbors, the powerful and deeply religious Egyptians. When we begin to look further afield, however, to regions of the world which these cultures are not thought to have influenced, we might be surprised to find similarities in places like the South Pacific. And yet that is exactly what we do find.

In the South Pacific island of Papua New Guinea, just north of Australia, there are over 850 tribes and cultures. One of these, the Angu or Kukukuku, a warlike tribe living in the high mountain ranges of Morobe, have a legend in which "they say two brothers killed an opossum and placed its bones in a stream. Gradually these grew into a man. Then one day the brothers found that the man had turned into a tree in which a bird sang. The brothers cut the tree down and the different clans came out singing. The women who were in the valley below came up to investigate the noise and were claimed as wives. The brothers named each couple and assigned them their own place to live" (cited in Roslyn Poignant's 1967 work, *Oceanic Mythology: the myths of Polynesia, Micronesia, Melanesia, and Australia*, 90).

Is it merely a coincidence that we find in this remote mountain tribe a slain opossum who turns into a man who turns into a tree, with the seemingly superfluous detail of a singing bird in its branches? It is obviously a special tree, because when it is cut down (!) the first representatives of the clans of the Kukukuku came out of it.

Similarly, on the tiny coral atoll of Ifaluk, in the Caroline Islands chain, there is a legend of two brothers who try to cut down a breadfruit tree, but it keeps going back together miraculously after they cut it down. At last, one of the brothers (the younger and more attentive brother) learns from his father that the bird sitting near the tree is really a god in disguise, which could be discerned because the bird "uttered the high trilling note that gods make" (Poignant 80). As with the myth of the Angu or Kukukuku society, the elements of the ancient myth of Isis and Osiris (a bird which is actually a deity in disguise around a sacred or miraculous tree) are faint, but they are distinctive enough to raise the possibility of some ancient contact between those societies (or contact between the forerunners of those two societies), long ages ago.

For the conventional theories, these manifestations of patterns found in long-vanished civilizations such as Egypt and Babylon among isolated islanders in the Pacific pose some problems. In the conventional anthropology, which consciously mirrors and reinforces the reigning Darwinian paradigm, man slowly inched forward from a "primitive hunter-gatherer" phase to an agricultural phase and then to greater specialization and technological advance. The common understanding is that some areas (particularly around Europe) advanced faster than others, and then colonized and oppressed those cultures that had remained in the hunter-gatherer mode for longer (the American Indians, the tribes of Africa, etc). An entire theory of critical analysis, "post-colonialism," is built upon these assumptions.

But these echoes of ancient mythology among isolated islanders raise another possibility: that they are also the descendents of an ancient and highly advanced civilization (or civilizations). In some isolated enclaves, some of this knowledge was lost or forgotten (as much of it was apparently lost or forgotten in many other cultures as well). However, because of their long isolation and remote security from the forces which stamped out such knowledge elsewhere, the people of the far-flung islands of the Pacific may preserve many aspects of things once known. For one thing, their navigational ability and knowledge of the stars, which they used in order to cross the vast Pacific with confidence and without compasses, prompted de Santillana and von Dechend to call the peoples of this region "the greatest navigators our globe has ever seen" (437). They preserved extensive knowledge of the stars, and the beautiful Pleiades held special significance in the annual cycle (located near Taurus and thus not far from Orion, the Pleiades have a similar cycle of visibility throughout the year). And, while Europeans of the eighteenth and nineteenth centuries looked down upon their methods of sustenance as "primitive," and it is true that agriculture may not have been as broad or as extensive in the islands than on the continents, it is also true that the first European visitors to Rapa Nui (Easter Island) reported extensive kumara plantations.

It is certainly true that in some places cannibalism also became part of the culture, which suggests a return to barbarism (the culture-hero associated with Osiris in many cultures is described as teaching men to grow food and to stop eating one another). It is quite possible that in some circumstances men finding themselves on isolated islands with fewer resources to go around (particularly with fewer marriageable women) might have erupted in violent conflict, as is known to have happened among the men from the HMS *Bounty* and from Tahiti on Pitcairn, and as apparently happened between different factions or tribes on Rapa Nui (Easter Island) at one point. In each of these two known historical cases, all or almost all of the men from one side were slaughtered. If most – or all – of the adult men were killed in such internecine combat, the children could certainly grow into adulthood without having the technological and astronomical knowledge passed onto them, or with only an incomplete understanding of what they would have otherwise learned. They would then grow into old age without the understanding that their forefathers had possessed, and being unable to pass it on to their children, such knowledge would vanish, leaving only an echo in the myths or religions of their descendents.

This possibility is at least as strong as some of the other theories that have been put forth to explain these mysterious mythological echoes. In the past century, theories such as Freud's theory that common psychological patterns found in most families are responsible for the similarity of these motifs around the world, and Jung's theories of a "collective unconscious" shared by all humans no matter how culturally removed, gained credence among scholars. Decades after the novelty of such theories has worn off, however, much of what was written by scholars who saw Freud and Jung as the solution to all analysis is being re-evaluated, and much of it looks somewhat silly in the perspective of history.

The suggestion that the common Freudian motifs arising from the upbringing of a child (his theory that tensions between the father, the mother and the child give rise to all the recurring patterns in myths around the world) or the suggestion that they arise from a Jungian "collective unconscious" appear to fall rather short in the face of the evidence we have examined so far. These theories certainly do not explain the presence of precessional numbers in the far-flung mythologies, unless Freud or Jung can somehow explain why the number 72 should be embedded in man's collective unconscious, or arise due to the family dynamics of a child's formative few years.

Of course, the alternative theory offered above (that such mythological echoes are not manifestations of some collective unconscious that sprang up among peoples who had absolutely no contact with one another, but are rather connected to the remnants of the knowledge of an ancient and mighty world-traveling people, just as other such remnants found elsewhere are) might not be a very flattering or palatable theory among those whose reputations or identities are bound up in the dominant paradigm, and it will be railed against by scholars whose entire livelihood is built upon "post-colonialist" theory or authorship. But in the investigation of a mystery, all clues should be examined, and all possible explanations for the evidence should at least be entertained.

Another important clue to this mystery is the fact that, as de Santillana and von Dechend observed, the "chopping down of the tree" myth is very frequently associated with flooding, a flood, rising waters, and world destruction. In *Hamlet's Mill*, they quote from Sir James George Frazer's (author of the previously cited *Golden Bough*) account of the mythology of the Ackawois (a Carib people of Guyana, situated along the Atlantic in the northeastern lobe of South America) which he first published in his 1918 work, *Folk-lore in the Old Testament*:

> The Ackawois of British Guiana say that in the beginning of the world the great spirit Makonaima created birds and beasts and set his son Sigu to rule over them. Moreover, he caused to spring from the earth a great and very wonderful tree, which bore a different kind of fruit on each of its branches, while round its trunk bananas, plantains, cassava, maize, and corn of all kinds grew in profusion; yams, too, clustered round its roots; and in short all the plants now cultivated on earth flourished in the greatest abundance on or about or under that marvellous tree.

In order to diffuse the benefits of the tree all over the world, Sigu resolved to cut it down and plant slips and seeds of it everywhere, and this he did with the help of all the beasts and birds, all except the brown monkey, who, being both lazy and mischievous, refused to assist in the great work of transplantation. So to keep him out of mischief Sigu set the animal to fetch water from the stream in a basket of open-work, calculating that the task would occupy his misdirected energies for some time to come.

In the meantime, proceeding with the labour of felling the miraculous tree, he discovered that the stump was hollow and full of water in which the fry of every sort of fresh-water fish was swimming about. The benevolent Sigu determined to stock all the rivers and lakes on earth with the fry on so liberal a scale that every sort of fish should swarm in every water.

But this generous intention was unexpectedly frustrated. For water in the cavity, being connected with the great reservoir somewhere in the bowels of the earth, began to overflow; and to arrest the rising flood Sigu covered the stump with a closely woven basket. This had the desired effect. But unfortunately the brown monkey, tired of his fruitless task, stealthily returned, and his curiosity being aroused by the sight of the basket turned upside down, he imagined that it must conceal something good to eat. So he cautiously lifted it and peeped beneath, and out poured the flood, sweeping the monkey himself away and inundating the whole land. Gathering the rest of the animals together Sigu led them to the highest points of the country, where grew some tall coconut-palms. 263-264.

De Santillana and von Dechend also make note of Clyde E. Keeler's compilation of Cuna (or Kuna, as it is often spelled today) mythology, from the Cuna tribes who live in the San Blas islands on the Atlantic side of Panama, a compilation entitled *Secrets of the Cuna Earthmother* (1960). Keeler notes that "the Cunas have the Water of Life issuing from the Tree of Life when it is cut down" (64). Keeler then points out that the people of New Hebrides (in

Oceania, far to the west, north of New Zealand and northeast of Australia, now known as Vanuatu) believed the flood was released when "Zat felled one of the tallest trees (Tree of Life) in the forest and built an enormous dugout" (64). Similarly, he notes that the Pueblo Indians of New Mexico, Arizona, and Texas believed that a flood was released "when a sacred squash vine was severed" (66), while the Teuq Indians of Colorado believed that "the Maize Stalk (the Tree of Life to many Indian tribes) expanded to become an enormous canoe" in which a man and woman saved a pair of all living things (66).

To be sure, the authors of *Hamlet's Mill* are convinced that all these flood images have celestial meanings. They note that the celestial "globe" was divided by the celestial equator, and that the ancients described everything "below" that line (that is to say, the southern celestial hemisphere) as the waters or the deep. Thus, the constellations of the zodiac, which spends half the year above (on the north-pole-side of) the celestial equator and the other half below it, are divided into wet and dry.

A precessional explanation for the connection could be that the "unhinging" of the central axis (the felling of the central tree, which mythologically encodes man's perception that the axis is moving and not steadfast, due to precession) is connected with the changing ages (from the Age of Taurus to the Age of Aries to the Age of Pisces and so on). The changing of these ages is metaphorically described as the "destruction of worlds" – the collapse of the reign of the four previous constellations that anchor the year at the spring equinox, summer solstice, autumnal equinox and winter solstice, and their replacement with new constellations. Hence, the chopping down of a tree (displacement of the axis) is connected with the destruction of the world and replacement with a new world (passing of one celestial age and replacement with another). As de Santillana and von Dechend also point out, this same motif is graphically described in the Norse myths with the account of Ragnarokk, the doomsday of the gods, in which the great tree Yggdrasil is split from top to bottom, and the sun and moon and gods are devoured, to be replaced by a new heavens and new earth. (Not mentioned by de Santillana and von Dechend is the fact that, like other axis-trees the world over which have a divine bird nearby, the mighty world-tree Yggdrasil is always described as having a supernatural eagle perched in its uppermost branches).

As satisfying as this celestial and metaphorical explanation may be for the rising of the waters (the tilting of the celestial axis to expose the southern or "wet" portion of the celestial sphere), it is also possible that the celestial explanation also reflects a geological explanation. If the heavens as we see them now were actually brought into their current configuration by a cataclysmic event accompanied by an actual geological flood, then this could also explain the connection in countless ancient myths between a flood and the skewing of the celestial axis. Such a connection would not negate the celestial connection – on the contrary, the two phenomena would be connected. The knocking off-kilter of the axial spin of the earth could have been caused by a cataclysm that released a flood, bringing the heavens into the view we have today, in which the slow forward movement of the equinoxes and solstices passes through the zodiac constellations that we have been discussing and which fill the myths from around the globe.

As a matter of fact, the theory of Walt Brown comports perfectly with this explanation – or, put another way, this additional evidence from the remaining lore of mankind's ancient past, found around the globe, supports the thesis that Walt Brown put forward based upon his examination of the geological clues on the surface of the earth and under the sea. His theory posits the cataclysmic eruption of waters trapped under the crust, floodwaters which erupted with such violence that they both caused a cataclysmic flood and also eroded the sides of the continental plates, which enabled the basement rock beneath to spring upwards, initiating a sliding of the plates away from the rupture, until friction and the forces of physics caused them to grind and crash to a halt, pushing up mountains in the same way that the front of a car or truck will buckle and form ridges if it is driven into a wall or a tree. According to his theory, the highest of these – the Himalayas – so changed the weight distribution of the spinning earth that the entire globe rolled. There is plenty of evidence that this did in fact take place – including fossils in the region of the Arctic circle of plants and animals that must have lived in temperate latitudes, and similar fossils in and around Antarctica. If so, the sky would have been forever altered. An entirely new set of stars and constellations would have been located at the celestial north pole (and celestial south pole). If this actually took place during human memory, it is not at all surprising that ancient people associated the unhinging of the axis with the flood, an unhinging which is often described in myths that appear to refer to the phenomenon of precession.

Writing long before Dr. Brown worked out his theory, Keeler in the *Secrets of the Cuna Earthmother* notes this about the nearly-universal flood accounts in South America: "Most South American stories speak of the Flood as being due to breaking of the great water jars of the underworld by a god, which is the Cuna method of producing rain" (60). Isn't it interesting that the fundamental aspect of Brown's theory, that the waters erupted from under the earth, receives reinforcement from the mythological evidence handed down from people who retained their ancient heritage all the way up into the twentieth century?

Another important observation put forth by de Santillana and von Dechend is the association of the central "tree" with a whirlpool, and with the abode of the dead. It is perhaps not surprising that the central axis be associated with a whirlpool: if one considers the celestial north pole during a single night, for example, it will be clear that the stars and constellations rotate around it during the night in a manner very suggestive of a whirlpool. The illustration in the chapter on heavenly phenomena on page 48 showed the movement of the Big Dipper around the celestial pole which is caused by the daily rotation of the earth, very suggestive of an analogy of a whirlpool. This leads me to conclude that the whirlpools of ancient myth (so often situated near a tree) refersto the celestial north pole, with the tree representing the axis of heaven. Mythical heros often have a dramatic encounter with this whirlpool and tree (or vine), since these myths encode celestial facts.

Most famous of the whirlpools of ancient mythology, perhaps, is the whirlpool of the monster Charybdis, encountered by Odysseus in the *Odyssey* and described by Circe in Book XII of Homer's epic. Here are the relevant passages, from the 1919 translation by Dr. A. T. Murray:

> Then queenly Circe spoke to me and said
> "All these things have thus found an end;
> But do thou hearken as I shall tell thee,
> and a god shall himself bring it to thy mind.
> To the Sirens first shalt thou come,
> who beguile all men whosoever comes to them.
> Whoso in ignorance draws near to them
> and hears the Sirens' voice, he nevermore returns,
> that his wife and little children may stand at his side rejoicing,
> but the Sirens beguile him with their clear-toned song,
> as they sit in the meadow, and about them is a great heap
> of bones of mouldering men, and round the bones
> the skin is shriveling. XII.39 – 52.

In the superlative translation of the late Robert Fagles (1933 - 2008), the line about the Sirens and "their clear-toned song" is translated thus: "The high, thrilling song of the Sirens will transfix him." Could there be any connection here between the "high, thrilling song" of the Sirens and the "high trilling note that the gods make" in the myth of the Ifaluk people of the Caroline Islands?

It is clear from this passage in the *Odyssey* that the trial of the Sirens and the trial of Scylla and Charybdis are connected, for as soon as Circe explains how Odysseus can have himself tied to the mast in order to enjoy the irresistible song of the Sirens, she describes the horrible choice between those two female monsters. Further, the description of the whirlpool of Charybdis is also connected to the account of the Clashing Rocks, which we earlier argued could well be a mythological analogy for the shifting equinoxes – as we will see when Circe continues (this time from the 1900 translation by Robert Butler):

> When your crew have taken you past these Sirens, I cannot give you coherent directions as to which of two courses you are to take; I will lay the two alternatives before you, and you must consider them for yourself. On the one hand there are some overhanging rocks against which the deep blue waves of Amphitrite beat with terrific fury; the blessed gods call these rocks the Wanderers [Fagles here translates: "the Clashing Rocks they're called by all the blissful gods"]. Here not even a bird may pass, no, not even the timid doves that bring ambrosia to Father Jove [. . .] XII.61 – 70.

Once all these significant surroundings are described, Circe goes on to describe the whirlpool itself, and the significant tree (in this case, a fig tree)

> Of these two rocks the one reaches heaven and its peak is lost in a dark cloud. This never leaves it, so that the top is never clear not even in summer and early autumn. No man though he had twenty hands and twenty feet could get a foothold on it and climb it, for it runs sheer up, as smooth as though it had been polished. XII.81 – 88.

She describes Scylla, whose six heads make a yelping voice no louder than a "young hound" in her cave "so high up that not even the stoutest archer could send an arrow into it," but terrifying in their ability to sweep the ocean beneath of any "dolphins or dogfish or any larger monster that she can catch." At last, on the other side of Scylla's rock, we meet Charybdis:

> You will find the other rocks lie lower, but they are so close together that it is no more than a bow-shot between them [Fagles translates this as "an arrow-shot apart"].

De Santillana and von Dechend spend some time on the celestial significance of arrows elsewhere in mythology, some of it involving whirlpools, and it is notable that arrow-shots are mentioned in conjunction with dog-headed Scylla as well; one possible connection de Santillana and von Dechend make is to the bow-and-arrow shaped group of stars which point to the all-important "dog-star" of Sirius, one of the most mythologically-connected stars in the heavens. Circe continues describing rock:

> A large fig tree in full leaf grows upon it, and under it lies the sucking whirlpool of Charybdis. Three times in the day does she vomit forth her waters, and three times sucks them down again; see that you be not there when she is sucking, for if you are, Neptune himself could not save you.

Note the almost "throw-away" line, but very significant, that this whirlpool Charybdis is more powerful in some way than even Poseidon, the god of the sea and of earthquakes, elder brother of Zeus – or at least, those sucked down her vortex are beyond even his aid.

> You must hug the Scylla side and drive the ship by as fast
> as you can, for you had better lose six men than your whole
> crew. XII.111 – 121.

Odysseus follows her advice, and six of his stoutest crewmen are
seized and devoured by the writhing heads of Scylla before the
ship can sail past to the island of the Sun (is it not significant that
these terrifying "gates" are all closely associated with the abode of
the sun?). There, disaster strikes, the main cause of much of Odysseus' wandering and suffering, according to his account. Afterwards, the South wind blows him back "toward cruel Charybdis,"
a trial he reaches for the second time at break of day after a night
of being blown by the gale. This time, his ship is sucked down
into the abyss:

> I was borne along by the waves all night, and by sunrise
> had reached the rock of Scylla, and the whirlpool. She was
> then sucking down the salt sea water, but I was carried aloft
> toward the fig tree, which I caught hold of and clung on to
> like a bat. I could not plant my feet anywhere so as to stand
> securely, for the roots were a long way off and the boughs
> that overshadowed the whole pool were too high, too vast,
> and too far apart for me to reach them; so I hung patiently
> on, waiting till the pool should discharge my mast and raft
> again -- and a very long while it seemed. A juryman is not
> more glad to get home to supper, after having been long
> detained in court by troublesome cases, than I was to see my
> raft beginning to work its way out of the whirlpool again.
> At last I let go with my hands and feet, and fell heavily into
> the sea [. . .] XII.464 – 478.

De Santillana and von Dechend point out other whirlpools in
myth, such as the "whirlpool island" in the legends of Borneo (the
third-largest island in the world, located north of Australia and
west of New Guinea) a whirlpool that is right next to a miraculous
tree one could ascend to the stars, to visit the "land of the Pleiades"
(213).

In the Kavya literature of ancient India, written by Jaina, Buddhist,
Vedic, and Saiva authors, the tale of Satyavrata also features a
whirlpool and a fig-tree, as well as a mysterious mountain, as

described by Anthony Kennedy Warder (in *Indian Kavya Literature*, vol 6, 1992):

> In the morning, Satyavrata declared that he would take Saktideva to the island of Ratnakuta, where Ocean had established an excursion festival of Hari in the month of Asadha. As people came for it from all the islands, he might learn from someone about the Golden City. On the way, Saktideva saw something rising from the sea, like a mountain with wings, which Satyavrata, who was at the helm, said was a fig tree, beneath which was a whirlpool which they must avoid. But the current carried the ship towards it and Satyavrata declared they were lost, but that Saktideva should be able to hang on to a branch of the tree, whilst he checked the ship a little, and might find a means to survive. This manoeuvre succeeded but the heroic Satyavrata bringing about the advantage of another lost his own life, with his ship, drawn down into the submarine fire. Clinging to a branch, Saktideva still in despair at sunset saw many great birds coming to the tree for the night. He overheard their talk of places they had visited, as they settled among the branches, and one old bird said he would go again to the Golden City in the morning to feed. During the night Saktideva hid under the feathers on that bird's back as it slept. In the morning the bird flew to the Golden City and alighted in a park, where Saktideva slipped down quietly from its back and got away. 579 - 580.

The similarities to the account in the *Odyssey* are striking, and it would be difficult to argue that they did not share a source. To come up with some way to explain the commonality of the fig tree above the whirlpool which swallows the ship, scholars must posit some ancient contact or transmission between Greece and India, or between cultures that took the story from one to another, or to both originally. Given the geographic proximity of the two locations, and the lack of a deepwater ocean between them, such a transmission does not appear too unreasonable, even within conventional paradigms.

Much further away and much more difficult to explain are manifestations of the same pattern in the Pacific islands.   Martha Beckwith, in her exhaustive study of Polynesian legend, *Hawaiian Mythology* (1940), recounts the Hawaiian romance of the hero Kae and the Island of Women (Vainoki or Vaino'i) and its variations among other peoples of Oceania, including the islands of the Marquesas, Rarotonga, Rapa Nui, and others. We might observe as an aside that, just as Kae in the South Pacific encounters an Island of Women, the *Odyssey* could equally be seen as an encounter with not one island of women but rather an endless series of "islands of women," among them the islands of Calypso, Circe, the Sirens, Scylla and Charybdis, the Phaeacians, and of course Ithaca itself and the trial offered by Penelope to her long-absent husband at the end. So, an alert student of mythology might sense already a connection between the legends of Kae and Odysseus.

In the Maori variation, the hero "joins Whiro's canoe party but when it enters a whirlpool he catches the overhanging boughs of a tree and lives among the Nuku-mai-tore, to whom he teaches the use of fire, the art of cooking, and the natural way of childbirth, together with the ceremonies attending the natural birth of a child" (502).

Beckwith, who studied a vast collection of accounts from the Pacific islanders, notes that: "The Kae story is not always connected with the teaching of natural childbirth and the use of cooked food. Interest is sometimes centered upon the whale-brother carrier whom Kae has cut up and eaten, whereupon avengers are sent who pack Kae into a canoe (or basket) in his sleep and bring him back to be killed (and eaten). The Maori say that with Kae cannibalism began" (504).

While the details are a bit disjointed, these connections bring to mind the slaying of a brother and cutting him into pieces (Seth and Osiris) and the imprisonment of the brother in a casket that is placed upon the water (as Osiris was, similar to the canoe or basket described above). Thus, this Maori version of the story of Kae who, like Odysseus, escapes the whirlpool by clutching a nearby tree also points to the precessional mythology of Set and Osiris, and indicates the connection between the world-tree (the Djed-pillar that encased the body of the slain brother) and the tree at the edge of the whirlpool.

The "tree which encloses the sky," then – to cite the ode to the Djed pillar found in the ancient Egyptian Pyramid Texts – is that celestial pillar about which all the heavens turn, like a great whirlpool. In many mythologies of the world, this central tree is chopped down: recall the Epic of Gilgamesh and the celestial Cedar Forest, or the mighty world-tree Yggdrasil of Norse legend, as well as the Cuna legend of Sigu and the marvelous tree, or the tree Zat cut down to initiate a flood in the legends of Vanuatu.

The association with flooding may provide a clue that a cataclysmic flood was somehow connected with a "chopping down" or alteration of the earth's axis, or the alteration of the sky that may have been associated with the initiation of its precessional rotation. The abundance of these startlingly similar "world-tree" motifs in cultures that conventional anthropology teaches were never in contact does not necessarily indicate an understanding of precession in all these peoples, although the portrayal of a Djed-like pillar in the "Seth and Horus" scene found not only in ancient Egypt but in Central America and in India hints that such knowledge may well have been present across the globe, and it is certainly likely that at least a cadre or priesthood knew the real significance behind the legends in these various cultures.

The fact that the flood and the unsettling of the axis are connected in the legends around the world provides evidence that these two terrifying events were in fact connected, and took place within human memory – just as the hydroplate theory of Walt Brown maintains.

Certainly, the striking similarities indicate the strong possibility of transoceanic contact in ancient times. The only other possibility is that such details just cropped up independently among people who had no contact at all (a suggestion that borders on the absurd).

The evidence from myth that we have examined thus far raises king-sized problems for adherents of the conventional paradigm. In the next chapter, in which we examine the precessional numbers found in myths, the problems become even greater and the case for an alternative explanation even more compelling.

The Precessional Numbers

We have seen extensive evidence that the mythology of Osiris, Seth (or Set), Horus and Isis encodes the celestial phenomena caused by the precession of earth's axis, which slowly over the course of many centuries displaced Orion as the constellation dominating the sky over the sun's rising on the day of the spring equinox. We saw evidence in the stars for the origins of the conflict between Seth and Horus, and for the mythological symbols which represent the axis around which the entire sky turns, an axis which the ancients perceived to have been unhinged, the unhinging of which is bound up with precession itself. Among the evidence we have examined, we briefly noted that Plutarch's account of the Osiris mythology contains the significant number 72 as the number of henchmen of Set who conspired in the original ambush of Osiris. The significance of this number, when found in a legend that involves the precession of earth's axis, lies in the fact that it is the closest whole number to the actual number of years required for the precessional circle to advance one degree.

This detail is an important piece of supporting evidence for the argument that this central theme of ancient Egyptian mythology – a theme that was somehow distributed around the world and which pops up in the legends of people living in the most unlikely places – is in fact precessional in its nature. It is also a detail with staggering implications for the conventional theorists, a detail which shakes their assumptions to their foundations.

First of all, as we have noted, conventional history (committed as it is to the narrative of man's slow advance through the millennia) teaches that it was not until late in the intellectual progress of the days of ancient Greece that man even conceived of the phenomenon of precession. Traditional textbooks will inform the reader with great certainty that it was not until Hipparchus of Nicea (who lived from 190 BC to 120 BC and whose work we discussed previously) that the locations of the stars were marked with enough precision to observe that they were drifting from the positions that they were supposed to occupy on any given day of the year. However, the evidence we have seen in the myths we have examined so far indicates that this knowledge of precession had been known long before, and then either lost or hidden from all but the initiated. Because it was apparently lost (or extremely well hidden) prior to Hipparchus' analysis, we should therefore take nothing away from Hipparchus and his achievement in rediscovering precession. It is nevertheless becoming clear that some ancient people had worked out the concept long before.

Still more unsettling for the conventional narrative, it appears that those ancient people had worked out a far more precise understanding than even Hipparchus (or his more famous intellectual and astronomical heir, Ptolemy) was able to achieve, with all the powers available to them at the height of ancient Greek and Roman technology. As we have already discussed, Hipparchus and Ptolemy after him both used one degree per hundred years as their general estimate of precessional advance, a number far less accurate than 72. Granted, Hipparchus knew that one degree every 100 years was the lower boundary for the rate, but there is no evidence that either he or Ptolemy were able to pin it down to the correct number of one degree every 72 years. This fact in itself argues that Plutarch or one of his contemporaries did not insert the precessional number 72 into Isis and Osiris narrative with knowledge available to them that was not transmitted from earlier centuries: to the contrary, it appears that the originators of the Osiris series knew more than those who followed them thousands of years later.

Further, the use of the number 72, in conjunction with precession, implies the use of a circle of 360 decrees of arc. If we can say that the earth's axis precesses one degree in approximately 72 years (and it does), then it is only because we are dividing a circle (in this case, the circle traced out by the axis in its precessional wobble) into something we call "degrees," which we have defined as 1/360ᵗʰ of a circle. If we were defining a degree, say, as 1/100ᵗʰ of a circle, then 72 would no longer be a precessional number, since the number of years to precess a "degree" of that distance would be much longer (it would take about 259 years instead). Thus, the observation that precession is moving the starry background by about one degree in 72 years cannot predate the division of circles into 360 degrees (in fact, it probably argues that such a division had been done even earlier, since dividing a circle into degrees is a much more basic mathematical advance than is observing the impact of precession upon the annual motion of the stars).

But why is the circle divided into 360 degrees, anyway? Certainly this is not an obvious choice. There are indications that the year itself was once divided by mankind into 360 days. The Egyptians, in fact, divided the year thusly, with the addition of five special days that were clearly separate from the original 360. In fact, as we have already noted, these five special days were "won" by the god Thoth in a game of draughts according to Egyptian myth, and they were such special days that the Egyptian mythology has them serving as the birthdates of the five important gods Osiris, Horus the Elder, Set, Nephtys, and Isis.

It is of course possible, as conventional theory avers, that early man (perhaps direct pre-dynastic predecessors of the Egyptians) mistakenly divided the year into 360 days, and later increased its precision by adding new days for greater accuracy. It is also possible that the year actually once contained only 360 days – in other words, that the earth once spun at a slightly slower rate, such that it only rotated 360 times in an annual circuit rather than its present 365.2420.

Could the change in the length of the year been caused by some cataclysmic event in earth's past? If so, it would not be surprising to find that the change was recorded in myth just as the phenomenon of precession is recorded in myth.

In fact, according to the hydroplate theory of Dr. Walt Brown, it is scientifically very possible that the same cataclysmic event which unhinged the axis and initiated precession also caused an increased rate of spin in the earth, both of which appear to be connected in ancient myth as well.

On page 107 of the seventh edition of his *In the Beginning: Scientific Evidence for Creation and the Flood*, Brown writes:

> The sudden formation of major mountains [during the "compression event" following the cataclysmic rupture, flood, rapid continental drift and then compression due to friction or collision, creating massive buckling] altered the spinning earth's balance, causing the earth to roll about 45°. The preflood North Pole moved to what is now central Asia. (That shift produced a 6° precession of the earth's axis that Dodwell discovered from studying almost 100 ancient astronomical measurements made over the last 4,000 years). This is why so much coal is found at today's South Pole and why many researchers have discovered lush vegetation, vast dinosaur remains, and frozen mammoths inside the Arctic Circle. These locations were at temperate latitudes before the flood.

> An impressive ancient record of a catastrophic flood and an apparent earth roll has been found. Charles Berlitz (1914 – 2003) reports that early Jesuit missionaries in China located a 4,320-volume work "compiled by Imperial Edict" and containing "all knowledge."

>> . . . *The earth was shaken to its foundations. The sky sank lower toward the north. The sun, moon, and stars changed their motions. The earth fell to pieces and the waters in its bosom rushed upward with violence and overflowed the earth. Man had rebelled against the high gods and the system of the Universe was in disorder.* [Berlitz, p. 126]

We will discuss the properties of physics which would cause such a roll in earth's axis. Before we do so, however, it is very revealing to note that the ancient Chinese record appears to describe flood-waters rising upwards from the "bosom" of the earth – in exactly the same way that the mythologies of the various indigenous peoples of South America, who (as we have seen) believed that the flood was caused by water upwelling from below, when "the great jars of the underworld" were broken, in conjunction with the felling of a great central world-tree image.

Even more noteworthy is the account of the early Jesuit missionaries of a multiple volume work, compiled by Imperial Edict to
contain all knowledge, that took up exactly 4,320 volumes. This
number – whether there actually was such an encyclopedia in so
many volumes or whether that number was only chosen symbolically – is a precessional number, as we shall see.

Dr. Brown goes on to explain the physics which would cause
earth's "big roll." Many "alternative" authors (as those who operate
outside of the orthodoxy of the accepted academic paradigm are
labeled) have posited a collision with some huge asteroid, comet
or planet (particularly Venus, which spins on its axis in the opposite direction from earth and the other planets) in earth's distant
past, but Brown argues that the known laws of physics do not
support this explanation, nor have proponents of such theories
been able to model the way in which such a collision would alter
earth's axial rotation.

On the other hand, during the compression event in which the
granite plates of the earth suddenly buckled like the hood of a car
in a head-on collision, several forces would immediately begin to
act on the spinning earth:

> As each mountain suddenly rose, its distance from earth's
> spin axis increased. This, in turn, increased each mountain's
> centrifugal force [. . .], a force that always acts away from
> and perpendicular to the spin axis. (A rock whirled at the
> end of a string produces an outward, or centrifugal, force
> that pulls the string taut.)

> Part of each new mountain's centrifugal force acted
> tangentially to the earth's surface and tended to roll the earth,
> so the mountain moved closer to the new equator. Because
> mountains are scattered around the earth, most of these
> "rolling" forces counterbalanced each other. However, the
> Himalayas and its plateau are so massive that their effect
> dominates all other mountains. (The world's ten highest
> peaks relative to sea level – including Mount Everest – are
> part of the Himalayas). In other words, the compression
> event created mountains whose centrifugal force rolled the
> earth so that the Himalayas moved toward today's equator.
> Also, the thickened, massive Eurasian hydroplate helped
> roll the globe in the same direction. 116.

We can think of the centrifugal force described above as the familiar force we would experience if we twirled a rock or weight around our heads, attached to a long string. The rock or weight would want to spin straight out – along the "equator," so to speak, of an invisible globe, with us at the center. It would be very difficult to twirl a rock without it going straight out – to twirl it so that the string made a sort of cone-shaped pattern instead of a straight-out circle. Similarly, the heavy weight of the Himalayas did not want to spin around a point in between the north pole and the equator – it wanted to spin around the equator, just like a rock on a string wants to spin straight out. The centrifugal force exerted by the Himalayas pulling towards the equator would act to roll the earth.

You could almost imagine the Himalayas as a huge handle that a giant would pull on towards the equator – this would act to move the spot that was formerly at the north pole "down" towards central Asia, and on the other side of the earth it would pull locations on the meridian opposite the Himalayas upwards, like chairs on a huge ferris wheel (in this case, the "ferris wheel" would be the meridian circle that runs roughly through the Himalayas and on around to the other side of the globe). This roll rotated Antarctica to the south pole, from a much lower latitude. The crustal-displacement theory attempts to explain the fossil evidence that shows Antarctica to have been temperate at one time by postulating the entire crust slipping around on the earth like the loose skin of an orange, but the hydroplate theory actually provides an explanation that would cause the entire earth to roll, not merely its "skin."

Interestingly enough, Dr. Brown explains that the earth's spinning orientation would remain pointed in the same direction in space – it would not move along with the "land" of the old north pole when that land rotated to a new latitude. The earth would remain spinning, but there would be new "land" underneath the point of spin in the north and the south (of course, in the north pole, it is not land anymore, but sea that is underneath the spinning point of the earth).

While this fact may seem counterintuitive, it is actually an example of the conservation of angular momentum, which says that a spinning body's spin axis does not change unless an external torque acts upon it. High school and college physics teachers often illustrate angular momentum with a large bicycle wheel. If a bicycle wheel is held in the air such that it is vertical (not horizontal, but aligned as it would be on a bicycle), and a string is attached to the axle on one side, we would expect it to fall over when released and end up lying horizontal, suspended from the string. And so it would, unless the wheel is vigorously spun by the teacher prior to releasing it, in which case it will acquire angular momentum that will keep it vertical, to the amazement of the students seeing it for the first time (this power of angular momentum conservation is somewhat counterintuitive when first observed until it is understood, just as the fact that the spinning earth would not move its spin axis even as it underwent its "Big Roll" is counterintuitive unless the principle of the conservation of angular momentum is understood).

However, Dr. Brown's theory not only offers a cause for this Big Roll (which fossil evidence supports), but explains that this Big Roll would have initiated a torque vector from the inner earth, which would have tried to keep spinning in its original orientation when the Himalayas were trying to rotate outward to the equator. This friction would have been counter to the direction that the earth was trying to roll, and would have had an impact on the obliquity of the ecliptic, temporarily increasing it by almost a full degree. Over time, the obliquity would return to equilibrium at about the angle it is today.

This explanation of a force that would temporarily increase earth's axial tilt would explain the findings of George F. Dodwell, who is mentioned in the above-cited paragraphs from Brown's book. George Frederick Dodwell (1879 – 1963) was a Fellow of the Royal Astronomical Society of Australia and was appointed Government Astronomer for South Australia in June, 1909, a position he held for the next forty-three years, retiring on October 31, 1952. In 1933, he obtained a copy of a medieval manuscript of Belgian astronomer Godefroid Wendelin (1580 – 1667), which had been lost for three centuries and rediscovered in 1933 in the Library of the city of Bruges (Dutch *Brugge*).

That Dutch manuscript, as Dodwell explains, "contained, amongst other things, a list of ancient observations of the obliquity of the ecliptic, made by Thales, about 558 BC; Eratosthenes, about 230 BC; in his later years, Hipparchus, 135 BC; Ptolemy, 126 BC; and by several mediaeval astronomers up to the time of Tycho Brahe, 1587 AD, and of Wendelin himself, 1616 AD; together with Wendelin's theory of the cause of the change which had taken place during the ages up to his own time" (Dodwell, *The Obliquity of the Ecliptic: Ancient, mediaeval, and modern observations on the obliquity of the Ecliptic, measuring the inclination of the earth's axis, in ancient times and up to the present*, 1963).

Dodwell (following Wendelin) observed that the ancient estimations of the obliquity of the ecliptic (a phrase which describes the same tilt of the axis we have been discussing, describing it from the perspective of an observer on earth, from which it seems that the ecliptic -- or path of the sun – has been tilted or made "oblique" because of the tilt of the axis) were slightly larger the further back in time he went.

The easy conventional explanation is that the ancient astronomers were simply wrong, but Dodwell describes the methodology of the ancient observers and argues that these methods were capable of producing fairly precise estimates of earth's axial tilt (or the ecliptic's obliquity from the point of view of an observer on earth). The method used by the great geographer Eratosthenes (c. 276 BC – 195 BC), who accurately estimated the circumference of the earth and created a system of longitude and latitude, was the method of observing the shadows of a gnomon throughout the year. Eratosthenes made detailed records of his shadow measurements, and we still have at least some of his records.

Further, Dodwell argues that Thales of Miletus (c. 624 BC – 546 BC) very probably used the method of measuring the shadows of a gnomon as well, since he was "a kinsman, companion, and fellow townsman at Miletus" of Anaximander (c. 610 BC – 546 BC), who learned the method from the Chaldeans and introduced it to the Greeks. The brilliant Pythagoras (c. 570 BC – 495 BC) is said to have been a disciple of Thales (and Anaximander), and doubtless learned the method as well.

A gnomon in the northern hemisphere (at all latitudes north of the Tropic of Cancer but south of the Arctic Circle) will cast a shadow which points to the north throughout the year. The shadow will of course move throughout the day as the earth turns and the sun rises in the east (casting a long shadow early in the morning pointing to the west, but still north of the gnomon) and proceeds to the west (casting a long shadow late in the afternoon pointing to the east, but still north of the gnomon). If the tip of the gnomon's shadow is marked throughout the day, the markers will form a line north of the gnomon, running west to east, with north and south being indicated by the line drawn from this line's closest point to the gnomon directly to the gnomon itself.

As we saw from the diagrams in Chapter 3, the sun makes its most northerly track across the sky in the northern hemisphere on the summer solstice, the longest day of the year, because the earth's orbital path has brought it to the point where the axis is pointing towards the sun. Thus, the shadow-line cast by the gnomon will be closest to the gnomon on that date, since the sun (which is always south of the gnomon and the observer in the northern hemisphere north of the tropics) is furthest north or closest to the gnomon's top on that day. When the track of the sun is much further south, at its most southern point on the shortest day of the year (the winter solstice), the angle from the sun to the top of the gnomon will be much less steep, and the shadow line cast by the top of the gnomon throughout the day will be much farther north of the gnomon.

Using the angle to the gnomon from the shadow line, it is possible (using basic trigonometry and formulae known to Pythagoras for certain, and to Greeks who came after him) to calculate the angle of the sun's zenith on the two extremes, the summer solstice and the winter solstice. The difference between these two angles gives twice the angle of the tilt of the earth's axis (or, to use a different terminology, of the obliquity of the ecliptic).

The double angle of obliquity (the difference between the two angles of the sun at summer and winter solstice) was carefully recorded by the ancients, demonstrating that the axis of the earth was slightly more tilted the farther back in history we go. Dodwell found another argument against the theory that the ancients were simply wrong in their measurements (which is already unlikely, since the task of measuring the shadow of a gnomon is in no way difficult and the idea that ancients such as Thales, Pythagoras, and Hipparchus could not measure accurately is extremely farfetched). He noticed that the calculation of the angle of obliquity became progressively smaller as he progressed forward in history, and that if he plotted the measurements over time, the decrease in the angle of obliquity was a very satisfying fit with a logarithmic sine curve – in other words, the tilt in the earth's axis appeared to be greater in antiquity, and to approach its present value along a known curve to physicists, as would be expected if the axis had been upset at some date in the past.

Dodwell found that measurements spanning 4,000 years saw a decrease in axial tilt from 25° 10′ to the present tilt of 23° 27′. He posited some catastrophe occurring approximately 2345 BC which tilted the axis from a near-vertical orientation to a tilt of over 25°. As the earth recovered from that catastrophe, the axial tilt slowly recovered back towards vertical, reaching the present axial tilt and then remaining there in about the year AD 1850.

In compiling his arguments, he plotted the obliquity measurements of ancient Chinese solstitial observations taken at known latitudes and recorded by the official astronomers between the years 1100 BC and AD 1280. He describes the instructions in the *Chou Pei Soan King* ("sacred book of calculations called Chou-Pei" – Chou Pei perhaps meaning the "gnomon of Prince Chou") which emphasize "let there not be the very smallest difference between the degrees" of measurement, and mentions that the official carpenter was consulted for assistance in obtaining perfect measurements (which argues against the modern academic dismissal of ancient measurements).

This work records the estimates of the ancient Chinese in 1120 BC of the distance in degrees from the North Pole to the Tropic of Cancer, the distance from the North Pole to the Tropic of Capricorn, and by subtraction we can determine that the number of degrees between each of these tropics and the earth's equator were 24° 12′ 6″.

Note that they knew the earth was round, and understood the concept of the tropics (the furthest point that the sun's rays will be vertical to the north and south of the equator, on the June and December solstices, respectively), and they knew mathematically how to calculate those from the measurements of the sun's shadow and subsequent determination of the angle of the sun at its zenith at varying days of the year, long centuries before Pythagoras was born.

Dodwell also presents ancient Hindu measurements observed in Ceylon in 945 BC and in Hastina Pura (an ancient seat of government a few miles south of Delhi) in 510 BC; ancient Greek measurements recorded by Thales, Pythagoras, Pytheas of Massilia (an ancient navigator, explorer, and geographer who sailed north to Britain and northern Europe, described the geographic concept of the Arctic Circle, and is thought to have done so around 325 BC), Eratosthenes, Hipparchus, and Ptolemy; medieval Arabs and Persian measurements observed in Baghdad, Rakka, Damascus, Shiraz, Edessa, and Cairo between the years AD 820 and AD 1019; the measurements of European astronomers between AD 1460 and 1700 (including those of Wendelin and Johannes Kepler, 1571 – 1630); and hundreds of observations from France and England taken between 1660 and 1890.

He also extrapolates the axis from the existing ancient monuments at Karnak in Egypt, Stonehenge in England, and Tiahuanaco in South America (in Bolivia west of La Paz and near the border with Peru), which were aligned with the equinoxes and can be used to calculate the positions of the sun from those special days during their epochs and hence determine the tilt of earth's axis in those ancient times.

Dodwell's text, which was his life's work from 1933 until his death, is available online and the reader is encouraged to consult it for himself. The fact that the observations of antiquity follow a smooth curve towards the present obliquity is a strong argument in favor of the position that the ancient observers were not in error but that in fact earth's axis was upset at a point in time when men could observe the result, and that the obliquity slowly recovered until it reached its present angle about a hundred and sixty years ago.

Dr. Walt Brown's theory elegantly explains the evidence that earth underwent such a roll and that the axial tilt received some sort of shock in the past. The various alternative explanations (which usually involve some extraterrestrial body, such as the planet Venus, striking earth or passing so close as to seriously disrupt it) are not as scientifically consistent as the hydroplate explanation.

Brown writes: "Gravitational forces, such as the Sun, Moon, or planets acting on earth's equatorial bulge, cannot explain such a large and rapid change. Extraterrestrial bodies striking earth would provide a sudden change in axis orientation, not the pattern of decay Dodwell measured. Besides, for an impact to change the axis this much would require a massive and fast asteroid striking earth at a favorable glancing blow. The resulting pressure pulse would pass through the atmosphere and quickly kill most air-breathing animals – a recent extinction without evidence" (98).

Brown's explanation, then, is a scientific model based on physics, and one which provides an excellent explanation for a connection between the flood, the roll of the earth, the disruption of the heavens, and the disruption of the axis (or cutting down of the world-tree / Djed-pillar), which may have been understood by the ancients as connected with the flood as well.

Dr. Brown did not address the extensive mythology from around the world which supports this very explanation, by linking these same events: a chopping down of a tree or pillar associated with the central sky-axis, a great flood often described as rising up from below the earth's surface, and the phenomenon of precession. These clues from man's ancient past provide a powerful supporting argument to Dr. Brown's arguments, as well as to the arguments of George Dodwell. Far better than any collision with Venus, the mathematical progression in the recovery of earth's obliquity from an incident in the distant past to its present angle can best be explained by an incident such as the "compression event" described in Walt Brown's hydroplate theory, which created the massive Himalayas and initiated the Big Roll.

Leaving aside for now further inquiry into the question of what initiated the earth's axial tilt and its precession, we return to the presence of certain significant numbers in the mythology of

Significant Precessional Numbers

$$(72)(360) = 25{,}920$$

$$(12) \qquad (6) \quad (360) = 25{,}920$$

$$(12) \qquad (2{,}160) \ = 25{,}920$$

$$(6) \quad (2) \qquad (2{,}160) = 25{,}920$$

$$(6) \qquad (4{,}320) \qquad = 25{,}920$$

$$(6) \qquad (540) \ (8) \quad = 25{,}920$$

$$(6) \qquad (108) \ (5) \quad (8) = 25{,}920$$

he above table depicts one way of obtaining the most common-
-encountered precessional numbers found in ancient mythology.
egin with the approximate number of years in one degree of pre-
ession (71.6 years, rounded to 72). Multiply by 360 (the number
 degrees for a full circle) to obtain 25,920 years. Now, using the
ctors of 72, one can arrive at the other precessional numbers: 108,
16, 432, and 540. Note that factors of ten can be added or removed
om these, such that the number 2,160 or 43,200 are simply versions
 the precessional numbers 216 and 432.

the world as further evidence for the assertion that the ancients were very familiar with the principles of precession, and that this knowledge was distributed around the world. Jane Sellers points out the existence of these recurring numbers in *Death of Gods in Ancient Egypt*, and points out that the prolific authors and scholars of myth Joseph Campbell (1904 – 1987) and Robert Graves (1895 – 1985) had also observed these mysterious numbers (though without, as far as can be ascertained, connecting them with the phenomenon of precession). As well, De Santillana and von Dechend give numerous examples in *Hamlet's Mill* of these precessional numbers in ancient myths from around the world as further evidence for the existence of an ancient civilization which was able to traverse the globe and which possessed an understanding of precession and other astronomical knowledge millennia before conventional theories would allow.

To begin our investigation of these numbers, remember that Plutarch, in his account of Osiris, Set, and Isis, notes that Set had 72 companions in his plot to measure Osiris and shut him in a casket. We have looked at evidence that the entire foundation of the Osiris myth is the heavenly mechanism of precession, which slowly delayed the accustomed rising of Orion and Sirius (Osiris and Isis) in the pre-dawn eastern sky on the significant morning of the spring equinox. Thus, the connection of the number 72 with this myth (recorded by Plutarch, a writer living in a culture which did not even know that precession moves the sky by one degree every 71.6 years) is extremely revealing and strong support for the theory that a more ancient culture did know about precession and its rate, and that world myths including the Osiris series are transmitters of the knowledge of some ancient and now-forgotten civilization.

If the precession of earth's axis moves one degree in 71.6 years, then an entire circle of 360 degrees would take 25,776 years. If we round 71.6 to 72 (as effective transmission in mythological stories would require, since we can hardly expect to hear that Set imprisoned Osiris with the aid of 71.6 henchmen), then multiplying 72 by 360 gives 25,920 years for a complete return. If the sky is divided into twelve signs of the zodiac, and each "age" is defined (as we saw before) by a new sign dominating with its heliacal rise on the morning of the spring equinox (the Age of Taurus followed by the Age of Aries followed by the Age of Pisces and eventually the Age of Aquarius), then the division of this complete circle of 25,920 by 12 yields the number of years in an age, or 2,160 years (we could arrive at the same number by multiplying the number of years to precess one degree, 72, by one-twelfth of a complete circle of 360 degrees, or 30 degrees). Thus, 2,160 is also a "precessional number."

In fact, as the diagram above illustrates, we can obtain all of the prominent precessional numbers which pop up in the myths from around the globe through the simple process of breaking out the factors of the number 72 and the number 360 and expressing the number 25,920 as the product of different factors. We thus determine the following to be among the most important precessional numbers: 72, 108, 216, 432, 540, and 25,920. Multiples of these by ten (such as 2,160 for 216) are equally precessional in their significance, and myths often incorporate precessional numbers with one or more additional zeros at the end of each.

Among the myths cited by Sellers (as well as by de Santillana and von Dechend, and Joseph Campbell too) which include combinations of the above numbers are the Icelandic Poetic Edda (perhaps the most important extant compilation of Norse mythology), in which we learn that Valhalla's hall contains "five hundred doors and forty," through which on the last day "eight hundred fighters through each door fare / When to war with the Wolf they go" (191).

The warriors of Valhalla in Norse mythology, as most readers are aware, are slain heroes chosen by Odin and his Valkyries to feast in Valhalla until the battle of Ragnarok – the day in which the old sun and moon are swallowed up by Fenris the Wolf and the Midgard Serpent, and the old gods perish and Yggdrasil the world-tree splits. As in the Osiris series, then, these precessional numbers accompany an event with clear precessional implications. Here, sallying forth to battle, we have the precessional number 540 for the number of doors, times eight hundred warriors each, giving a product of 432,000 – which is of course another precessional number. Just as it is significant that the precessional 72 appears in a legend about a god who became a pillar that was cut down, it is significant that the precessional 540 and 432 appear in a legend about warriors sallying forth to the battle in which the old sun will be devoured and replaced by a new one.

Sellers also points out that Hindu texts dating from AD 400 describe a period of time known as a "yuga of the gods," made up of 12,000 periods of 360 years, each period of 360 years being known as a "year of the gods," and the number 12,000 being referred to as a "yuga" (Sellers 192). The product of 12,000 and 360, of course, is the significant number 4,320,000 – ten times the same product found in the description of Valhalla on the day of Ragnarok, and a version of the precessional number 432 (mythical transmission of these numbers often change them by factors of ten, or a hundred, or a thousand).

Recall that the early Jesuit missionaries in China reported the existence of a compilation of "all knowledge," compiled by Imperial edict, and contained in an encyclopedia in which there were 4,320 volumes. The explanation that these Jesuit missionaries were somehow experts in ancient mythology and made up the story of the Imperial encyclopedia (inserting a number they had found in an Icelandic Edda or an early Hindu text) seems far less likely than the more probable explanation that these same significant numbers were preserved in China as well as in Scandinavia and India, and in many other far-flung cultures around the globe.

It is noteworthy, also, that traditional Chinese martial arts often contain precessional numbers – for example, the complete Taoist Tai Chi set contains 108 movements, each with a descriptive name (number one: Opening of Tai Chi; number two: Left Grasp Bird's Tail; number three: Grasp Bird's Tail; number six: White Stork Spreads Wings; number eighteen: Carry Tiger to Mountain; number seventy-eight: Go back to ward off Monkey; number ninety: Move Hands like Clouds; number one hundred six: Appear to Close Entrance; number one hundred seven: Cross Hands; and number one hundred eight: Closing of Tai Chi). Other traditional Chinese martial arts descended from the Shaolin Temple also contain the number 108 in their forms. There is also the legend, repeated in many tales of the Shaolin temple, that in order to be deemed worthy to represent the Shaolin temple in the wide world, a monk would have to pass through the formidable Wooden Man Hall, in which were 108 wooden mannequins, each one designed to launch a different potentially deadly surprise attack on the candidate as he negotiated the test.

Since many of these martial arts were kept very secret and never taught to westerners until very recently, this is clear evidence that the precessional numbers were revered in China and preserved there, rather than having been imported by later contact with the west.

De Santillana and von Dechend note that the ancient Hindu Rigveda (one of the oldest of the Vedas, dating to before 1100 BC – and possibly several centuries earlier) contains 10,800 stanzas (another version of the precessional number 108) and that each stanza contains forty syllables: 432,000 syllables in all (162). However, it seems that this syllable count is attributed to the Rig by the Satapatha Brahmana (probably dating to a period between 700 BC and 500 BC in India) and does not actually match the count in the versions we have today (Thomas McEvilley, *The Shape of Ancient Thought: Comparative Studies in Greek and Indian Philosophies*, 77). Even so, the ancient assertion that the Rig contains this significant number of syllables predates the "discovery" of precession in Greece and Rome by many centuries, and whether the text contains exactly that many syllables is really immaterial. The fact that it is attested to have that many is what is significant. The Satapatha Brahmana contains other precessional numbers, asserting for instance that the stars in the sky number 10,800,000 (McEvilley 77).

De Santillana and von Dechend also point out that the ritual of the Agnicayana, which continues to this day in some parts of India and has been described as "the world's oldest surviving ritual" (having continued for over 3,000 years) involves the building of a brick altar using 10,800 bricks (162). Note that the Agnicayana altar is a furnace in the shape of a bird in honor of the Hindu god Agni. Agni is actually a title that is held by successive gods (passing like ages into the past, to be replaced by new deities who claim the title of Agni); traditionally there are four Agnis. The four Agnis are clearly associated with fire (often an indicator in myth of the equinoctial colure, as de Santillana and von Dechend point out elsewhere in their thesis) as well as with deep waters. One of the Agnis, de Santillana and von Dechend inform us, is named Apam Napat (*apam* "of the waters" and *napat* "nephew") – possibly a connection, de Santillana and von Dechend speculate, to the relationship of Seth and Horus (Seth the murderous uncle who slew Horus' father Osiris) and the Amlethus-myths used by Shakespeare as the basis for Hamlet, whose murderous uncle slew Hamlet's father (429-430).

The famous Maya long count was used for historic inscriptions and for calculating dates far into the past and the future. The system was elegant and simple, consisting of various periods of time which were indicated by their own glyphs; extremely large dates could be rendered by placing dots next to glyphs to indicate the numbers of the groupings (up to four, after which line strokes were used to indicate groups of five, with a zero symbol which could be used to indicate no units in that particular glyph-group). These groupings generally increased by factors of twenty (making the Maya long-count system a "base twenty" system, or vigesimal system, as opposed to our "base-ten" or decimal system of numbering).

The Maya *tun* contained 360 days, each day known as a *kin*. Twenty *tuns* was called a *katun* (and totaled 7,200 days). Twenty *katuns* was called a *baktun* (which would thus have 144,000 days). The long count cycle consisted of thirteen *baktuns*, an enormous length of time equal to 1,872,000 days (over 5,125 years).

As many are aware today thanks to the great interest surrounding the end of this great long count cycle, the Maya (whose civilization is thought to have started as early as 2000 BC but which reached its Classic period between AD 100 and AD 900) began their cycle with the year that we would call 3113 BC – "believed to be the year the Maya considered as marking the creation of the present world order" according to astronomer Edwin C. Krupp (born 1944) in his 1983 work *Echoes of the Ancient Skies: the Astronomy of Lost Civilizations* (185). The great cycle of thirteen *baktuns* that preoccupied the Maya astronomers and historians so many centuries ago will reach its conclusion near the end of our AD 2012.

Whether the end of this Maya long-count cycle was meant to purport a world-wide catastrophe (as many in popular culture have been told) or whether it was really a symbolic transition indicating a change in the heavens or of time is a matter of debate among experts. What is not a matter of debate is the fact that thirteen *baktuns* would contain 1,872,000 days. By a fascinating "coincidence," this number indicates an understanding of the approximate length of a complete precessional cycle (26,000) times the fundamental precessional number 72.

We have now seen these precessional numbers turn up in the central myths of the ancient Egyptians, which were created at some point long before the first of the Pyramid Texts, in the Norse mythology of Odin and Valhalla, in ancient Hindu texts and rituals, in ancient China and in the traditional Chinese martial arts, some of which were kept secret from westerners and have very old roots and Taoist and Buddhist connections, and now among the numbers of the ancient Maya. How can conventional academics continue to maintain with a straight face that precession was not known by the ancients until the time of Hipparchus?

And yet there are still more examples.

Berossos, an ancient astronomer and historian born around the period 340 BC to 323 BC in Babylon (long after the epoch of the Babylonian Empire but during the Hellenistic period contemporary to the conquest of that region by Alexander) and who compiled several works which are now lost but which were quoted by other ancient scholars, compiled a list of ancient Sumerian kings (many or all of whom may have been mythological) whose reigns sum up to 432,000 years (Jane Sellers, citing Joseph Campbell's work on Berossos, Sellers 188). She also notes that Berossos asserted that the number of years from Creation to a "Universal Catastrophe" numbered 2,160,000 (189). As we remember from our chart of precessional numbers, both 432 and 216 are prominent indicators of precessional code.

"Was Berossos, in 280 BC, transmitting secret numbers relating to changes brought about by precession?" Sellers asks. "This would be startling if true" (188).

An echo of the Maya belief in an extremely long cycle of time beginning in what we would call the fourth millennium BC is found in the ancient Hindu text of the *Surya Siddhantha*, which speaks of the current age of man as the fourth and most degenerate (a theme found around the world, including among the ancient Greeks, as well as in the Maya cosmology, which speaks of three previous "suns" which were each ended by a catastrophe, the current age being the "fourth sun").

The text of the of the *Surya Siddhantha* launches at its outset into a discussion of the division of time and various cosmological and mythological cycles. In the words of an English translation by Pundit Ba'pu' Deva Sa'stri and published by the Baptist Mission Press in 1861, we find that "the time containing twelve thousand years of the Gods is called a *Chaturyuga* (the aggregate of the four *yugas, Krita, Treta, Dwapara and Kali*). These four *yugas* including their *Sandhya* and *Sandhya'ns'a* contain 4,320,000 solar years" (3). Later, we learn that "In a great *yuga* each of the planets, the Sun, Mercury, Venus and the *Si'ghrochcha* (i.e. the farthest point from the centre of the earth in the orbit of each of the planets) of Mars, Saturn and Jupiter moving towards the east make 4,320,000 revolutions (about the earth)" (5) and that "There are 1,577,917,828 terrestrial days and 1,603,000,080 lunar days in a great yuga" (6) – both of these large numbers being divisible by 432 (and hence also by 72, which goes into 432 six times). The parenthetical notations are in the original translation – today, we would probably indicate these comments from the translator in brackets.

Robert Graves noticed the importance of the number 72 in the secret, sacred alphabets of ancient Wales and Ireland (with connections to many more far-ranging cultures as well), which he describes in his monumental study, *The White Goddess: a Historical Grammar of Poetic Myth* (1948).

He writes that in the Ogham alphabet, "the aggregate number of letter-strokes for the complete twenty-two letter alphabet is 72, a number constantly recurring in early myth and ritual; for 72 is the multiple of the nine, the number of lunar wisdom, and eight the number of solar increase" (251). He argues for a correlation between a year of five times seventy-two days with five intercalary days (which was the arrangement of the ancient Egyptians) and the five vowels, which acted both as "intercalaries" and as dividers of the year into five seasons of seventy-two days each.

Graves points to the ancient division of Ireland into five provinces, and to a reference within the early Middle Irish poetic cycle of the *Saltair na Rann* "in which a Heavenly City is described, with fifteen ramparts, eight gates, and seventy-two different kinds of fruit in the gardens enclosed" (281). In addition, he provides substantial evidence from the alphabets themselves and the ancient arrangements of the letters, which often parallel ancient "rectangular zodiacs." While not making the direct connection between the number 72 and the rate of precession, Graves finds numerous connections between the ancient mythologies and the "grammar of poetic myth" that he treats in his text, including connections between certain letters and corresponding gods and goddesses, and between certain letters and corresponding trees, flowers, vines, and their properties.

He *does* discuss the precession of the equinoxes, though without direct discussion of the precessional numbers, noting evidence from the ancient alphabets that the alphabets originated at a time when the sign of Gemini dominated the spring equinox – an age before the Age of Taurus. He writes:

> The original Zodiac, to judge from the out-of-date astronomical data quoted in a poem by Aratus, a Hellenistic Greek, was current in the late third millennium BC. But it is likely to have been first fixed at a time when the Sun rose in the Twins at the Spring equinox – the Shepherds' festival; in the Virgin who was generally identified with Ishtar, the Love-goddess, at the Summer solstice; in the Archer, identified with Nergal (Mars) and later with Cheiron the Centaur, at the Autumn equinox, the traditional season of the chase; in the resurrective Fish at the Winter solstice, the time of most rain. (It will be recalled that the solar hero Llew Llaw's transformations begin with a Fish at the Winter solstice.)

> The Zodiac signs were borrowed by the Egyptians at least as early as the sixteenth century BC, with certain alterations

– Scarab for Crab, Serpent for Scorpion, Mirror for He-goat, etc. – but by that time the phenomenon known as the precession of the equinoxes had already spoilt the original story. About every 2000 years the Sun rises in an earlier sign; so in 3800 BC the Bull began to push the Twins out of the House of the Spring Equinox, and initiated a period recalled by Virgil in his account of the Birth of Man:

> *The white bull with his gilded horns*
>
> *Opens the year . . .*

At the same time the tail of the Lion entered the Virgin's place at the Summer solstice – hence apparently the Goddess's subsequent title of 'Oura,' the Lion's Tail – and gradually the Lion's body followed, after which for a time she became leonine with a Virgin's head only. Similarly the Water-carrier succeeded the Fish at the Winter solstice – and provided the water to float the Spirit of the Year's cradle ark. 379 – 380.

Interestingly, if there have actually been three ages prior to the present age (the Age of Gemini, the Age of Taurus, and the Age of Aries prior to the present Age of Pisces which is nearly at a close), then we have a clear parallel to the Maya teaching that we are in the Fourth Sun, and the ancient Hindu teaching that we are in the Fourth Yuga.

In sum, mythology furnishes more than abundant evidence that ancient man knew of precession and that those who understood it somehow spread that knowledge worldwide. Remnants of the precessional numbers survive in global mythologies, evidence that is as compelling as that found in physical archaeological remains, and that is just as important as a clue to the mystery of man's ancient past. We will see that ancient archaeology often encodes these numbers as well, when we turn to the evidence left in monuments such as the Great Pyramid and Stonehenge, as well as in numerous other ancient sites that may not be as widely known.

Precessional Numbers and Astronomical Alignments in

the Monuments of Antiquity

Perhaps the most obvious place to start an examination of the archaeological clues from man's distant past which can help shed light on the accuracy of the hydroplate theory – and which can perhaps be best explained in light of it – are the pyramids of Giza.

Volumes have been written about the pyramids, and especially the Great Pyramid, enough to show that there is substantial evidence – actively suppressed by the guardians of conventional orthodoxy – to undermine the reigning academic paradigm at several points.

Robert Bauval and Graham Hancock do an excellent job of outlining many of these in their studies of the pyramids, particularly their joint book *Keeper of Genesis: A Quest for the Hidden Legacy of Mankind.* In that work, they detail several findings which could set the reigning orthodoxy on its ear but which have been conveniently "lost," discredited, ignored, or in which permission to continue has been abruptly removed by those with a vested interest in preserving the status quo.

Among the findings Bauval and Hancock relate in detail are the "relics" discovered in the sealed "air shafts" leading to the Queen's Chamber in the 1800s, one of which – a cedar plank – could have been carbon dated later in the twentieth century, had the British Museum in possession of the relics not lost it somehow. Another is the presence of a small iron plate discovered embedded below the surface of the south face of the pyramid near the exit point of the shaft leading from the King's Chamber.

Because the presence of iron in a monument built so long before the "iron age" would upset the succession of man's slow advance to the mastery of progressively harder metals which forms a corner-stone of the conventional description of man's progress, this iron plate has been dismissed out of hand as a hoax planted by Colonel Richard William Howard Vyse (1784 - 1853; born Richard William Vyse, he assumed the additional name Howard after his marriage to the heiress of Sir George Howard).

Charles Piazzi Smyth (1819 – 1900), in his 1864 work *Our Inheritance in the Great Pyramid*, explains the situation that prevailed in his day, one which is even more entrenched over 150 years later:

> Say the German archaeologists, and after them the ordinary antiquaries of most other nations too – man must have begun as an ignorant, helpless savage, little better than an ape, and then have risen to his present state of civilization by his own slow improvements only, on the simplest possible beginnings. Wherefore, when these specious philosophers find, perchance, an accumulation of various implements on opening some old tomb, the rough stone tools amongst them are voted the oldest; the polished stone, or bone, the next; then the gold, copper, or bronze, as being easily worked metals; and last of all the more difficult iron, if, indeed, its terrible rapidity of rusting away has allowed its presence to remain in any tangible form. That is the theory: and then, to convert it into fact, the objects are unblushingly arranged in Continental Museums, *in that order*, ticketed with dates corresponding therewith; and then used for proving beyond doubt that the history of man on earth began with an age of rough stone weapons; then came an age of smooth stone or bone; then the age of bronze; and finally that of iron, under which we are now living. 590, italics in original.

Piazzi Smyth points out that this desired storyline (desired, that is, by the proponents of a new modern philosophy) of man rising slowly from primitive savagery is at variance with an opposing theory: "that the Bible declares the savagism of man to have prevailed, not in primeval, but mediaeval times; and was the fruit of sin, idolatry, and willful *degradation* from the high level on which he had been first placed by his Creator" (590, italics in original).

Today we have had over a century and a half of the kind of conditioning that Piazzi Smyth describes above, in which virtually every museum anyone has visited and every textbook anyone has been assigned to read since grade school has confidently presented the timeline of man's slow progress through "stone age," "bronze age," and finally "iron age" (with the iron age traditionally dated no earlier than 1200 BC, over a thousand years after the accepted construction of the pyramids). And yet, as Piazzi Smyth argues, the Great Pyramid, which he declares "the oldest piece of architecture amongst men, so far from showing primeval savagism, is still the largest and best built to be seen anywhere; and with its component stones so grand, so hard, and yet so truly and exquisitely cut and squared to accurate mathematical forms and sizes, that the workmen *must* have had *iron* tools abundantly at their command" (591, italics in original).

Setting aside the discussion of whether the Great Pyramid is in fact older than any other structure in Egypt or elsewhere, as well as the discussion of whether its builders had iron tools or not, the fact is that the clues which could answer these questions appear to have been systematically suppressed or neglected since Piazzi Smyth's day. He himself points out as evidence:

> a large piece of wrought iron which Colonel Howard-Vyse had extracted out of the solid masonry of the Great Pyramid at the bottom of a deep hole he had been forcibly making in it. This remarkable curiosity had been presented to the British Museum, but was sadly neglected and undervalued as to its tremendous importance. True, it was only *one* piece; but one piece, even if it had been of less size than this one, preserved through 4,000 years by the strange accident of having got embedded in mortar between the stones, is quite enough to prove a principle; while the years elapsed since then are far more than enough to have oxydized and caused

to disappear any number whatever of its once companion
iron implements in any ordinary situations and exposures.
591, italics in original.

Bauval and Hancock describe in detail the circumstances
surrounding the discovery of this iron plate by workers employed
by Vyse in May of 1837. They also describe the examination in
1881 of the plate by Sir W. M. Flinders Petrie, who noticed in the
rust on the plate the cast of a marine fossil, proving that it had
rested "for ages beside a block of nummulitic limestone" (Bauval
and Hancock, 105), lending credence to the fact that it was genuine
(the fact that Vyse and his men had not noticed or called attention
to this tiny fossil imprint argues that it was not somehow forged
by them as part of a hoax).

Again, we cite this example not as part of a discussion of the age
of the pyramid or the question of whether its builders had access
to iron implements, but as a prime example of a theme that domi-
nates the examination of archaeological evidence – the theme
eloquently described above by Piazzi Smyth of the preconceptions of those who subscribe to a certain paradigm regarding the
timeline of man's history, who arrange all evidence to support
this preconception, and who "sadly neglect and undervalue" any
evidence to the contrary, no matter how striking.

The attitude of those obstructing any examination of evidence that
could undermine the conventional view is summed up well in a
criticism by a principal defender of the conventional paradigm,
leveled at a 1994 television documentary on the mysteries of the
Sphinx which dared to explore theories outside of the sanctioned
explanations: "It is evident that this John West represents nothing
but a continuation of the cultural invasion of Egypt's civilization"
(cited in Bauval and Hancock, 97). This attitude, that explanations
outside of orthodoxy somehow threaten the accomplishments of
one race or culture, is perhaps the biggest obstacle today to any
search for the truth. Not just in Egypt but around the globe we
will find that evidence which undermines the current paradigm
cannot be examined, because it might offend the sensitivities
of races or cultures which see any alternative explanations as a

threat – a "cultural invasion" of their past history, even if that past history as it has been told for the past hundred years or so turns out to be nothing more than a fable.

We heartily commend the work of Robert Bauval and Graham Hancock in their examination of the extensive evidence at Giza that there is far more to the story than the conventional paradigm dares to admit, even if we do not agree with every single one of their conclusions. The crucial observations from the pyramids and Sphinx of Giza for the argument presented in this book are more narrow in scope.

First, we observe that the pyramids and the Sphinx remain amazingly and precisely aligned not only with the earth's true cardinal directions but also with important stars. As Bauval and Hancock point out, and as was first noted in the early 1960s by Alexander Badawy (1913 – 1986) and Virginia Trimble, the shafts of the Great Pyramid are aligned so precisely with certain regions of the heavens that we can still, using computer programs that "wind back" the precessional clock to ancient epochs, determine which exact stars they were designed to point towards. Similarly, the sides of the Great Pyramid have long been known to align more precisely with earth's cardinal directions (true north, south, east and west) than many of the most modern buildings (including, as Bauval and Hancock point out, the Meridian Building at the Greenwich Observatory in London!)(Bauval and Hancock, 60).

Because the sides of the Great Pyramid are precisely aligned to the cardinal directions, the passageways point to true north and true south. Because of the turning of the earth, as we saw in earlier chapters, stars (including the sun) transit the sky from east to west and reach their zenith as they cross the meridian that runs along the inside of the imaginary celestial globe from true north to true south (running through the celestial north pole around which the sky appears to rotate). This highest point is called the meridian transit of the star, and the angle at which any given star crosses this meridian will stay the same throughout the year, changing only by the motion of precession over the centuries. As shown in the cross-sectional diagram of the Great Pyramid, below, the builders of the pyramid designed the shafts and passageways to point north and south, at significant angles that align with the meridian transits of important stars in the night sky.

This fact in and of itself is damning to the theory of continental drift as it is taught in the dominant paradigm of plate tectonics. If the continents have been drifting for eons upon the circulating magma of the mantle like ships upon a sea, it is impossible that such an ancient monument would have preserved any of these

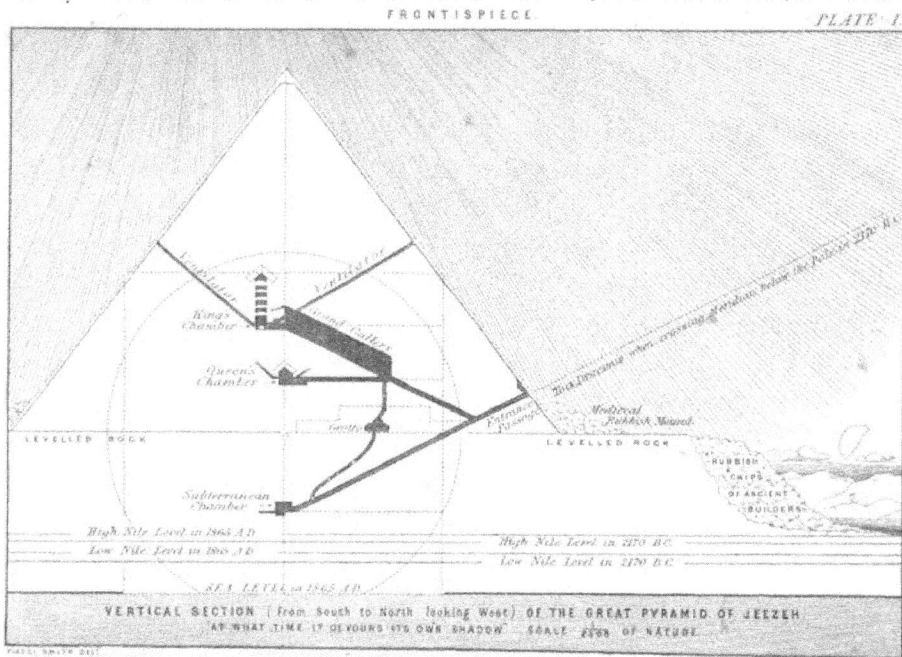

Diagram from the frontispiece of *Our Inheritance in the Great Pyramid, New and Enlarged Edition*, by Charles Piazzi Smyth (1819 - 1900), London: Isbister, 1874. Diagram depicts cross section of the Great Pyramid looking to the west, so that north points to the right of the viewer (viewer stands east of pyramid). Shafts leading Queen's Chamber were known but not explored, so they are drawn very short. We now know they penetrate at least 200 feet. The angles of all the shafts depicted could however be determined even then, and by 1871 it was known that the Entry Passage angle pointed to Thuban (*alpha Draconis*). In the 1960s, Alexander Badawy and Virginia Trimble determined that the King's Chamber shafts pointed to ancient meridian transits of Orion's belt in the south and Thuban in the north (the northern King's Chamber shaft ascends at the same angle as the Entrance Passageway).

extremely precise alignments. To repeat the statistics presented first in the introduction to this book, plate tectonics argues for a drift of approximately thirty millimeters per year (a little over an inch), and that even using the most conservative conventional dating for the Great Pyramid (which we believe is too recent, as will be discussed later) this would argue for movement of well over 380 feet in the ensuing centuries.

Another important aspect of the Great Pyramid is the measurement of its dimensions and the degree to which the pyramid (along with other ancient archaeological sites) incorporates the precessional numbers discussed in the previous chapter and found also in the mythology that serves as a parallel source of evidence from ancient history.

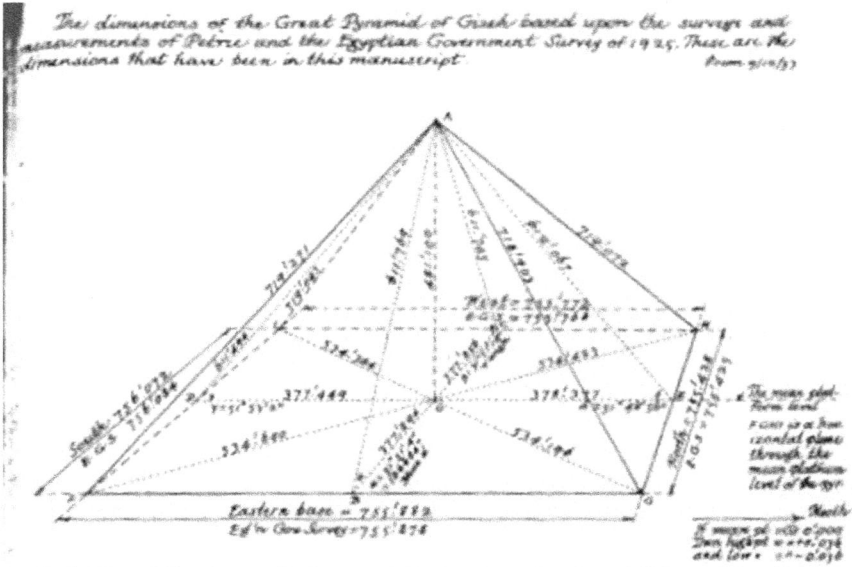

Diagram and dimensions of the Great Pyramid by J.H. Cole, published in *Determination of the Exact Size and Orientation of the Great Pyramid of Giza*, Cairo: Government Press, 1925.

While there may be some slight variance in the dimensions discovered from one technique and surveyor to another, the general dimensions of the Great Pyramid found by modern surveying methods are close enough to reveal the presence of a significant precessional number in the length of each side. As can be seen in the diagram above, from the 1925 pamphlet *Determination of the Exact Size and Orientation of the Great Pyramid of Giza*, using multiple theodolites and contemporary surveying techniques, the length measured for the four sides of the Great Pyramid are as follows:

East: 755.882 feet

West: 755.772 feet

North: 755.428 feet

South: 756.072 feet

Again, while other surveys may arrive at slightly different conclusions, these measurements will be sufficient to reveal the presence of precessional numbers in the base of the pyramid, which is powerful confirmation beyond the presence of 72 henchmen of Set in the Osiris myth related by Plutarch that the ancient civilizations possessed a refined understanding of the phenomenon of precession long before conventional theory accepts that they understood it.

These measurements, when converted to inches, yield the following:

East: 9,070.584"

West: 9,069.264"

North: 9,065.136"

South: 9,072.864"

In his 1999 book *Ancient Celtic New Zealand*, Martin Doutré argues for the existence of a royal cubit or "long cubit" of 21 inches, based on the arguments of Donald Lenzen in *Ancient Metrology*, who points to an ancient inscription at Siloam describing the distance covered by an aqueduct as 1,000 cubits and also 1,200 cubits. When the actual distance covered by the aqueduct is measured in modern units of measurement, we find that dividing the measurement by these two different cubit counts gives us a shorter cubit of about 18" and a longer cubit (or royal cubit) of 21" (135). Doutré later points out that the lengths of the sides of the Great Pyramid, if divided by the longer cubit of 21 inches, yield a length of 432 long cubits or royal cubits per side.

Using the measurements from the surveyor above and dividing by 21 inches per royal cubit, we would get the following:

East: 431.933 royal cubits

West: 431.869 royal cubits

North: 431.673 royal cubits

South: 432.041 royal cubits

In fact, a measurement of 9,072 inches divided by 21 would yield exactly 432. The inescapable fact is that the sides of the Great Pyramid appear to encode the number 432, the very same number (differing only by round factors of ten) found in the Norse Poetic Edda describing 840 warriors trooping through the 540 doors of Valhalla (for a total of 432,000). As seen in the previous chapter, 432 is an important precessional number, equal to 25,920 divided by 60 (or, to approach it from the other direction, six times 72, that very important round number of years for one degree of precession).

Many have previously noted that the pyramid itself relates to the dimensions of the globe of the earth on a scale of 1:43,200. This is remarkable in and of itself, in that conventional anthropology has a very difficult time explaining how an ancient civilization operating at least 2,600 years before Christ knew the precise measurements of our spherical earth! It is even more remarkable in that it indicates the ancient architects were also calling attention to the phenomenon of precession in the scale ratio that they selected for their model of our spherical earth.

If one adds the four side lengths from the 1925 survey cited above, the perimeter of the base is found to be 3,023.154 feet. The equatorial circumference of the earth is 24,901.55 miles (131,480,184 feet). If the pyramid perimeter is multiplied by 43,200 it yields 130,600,523 feet (24,734.9475 miles). As Graham Hancock and Robert Bauval point out, this is a result "that is within 170 miles of the true equatorial circumference of the earth" and that while "170 miles sounds like quite a lot, it amounts, in relation to the earth's total circumference, to a minus-error of only three quarters of a single per cent" (38). Hancock and Bauval argue the same 1:43,200 ratio between the height of the Great Pyramid and the polar radius of the earth. However, this is more difficult to prove, since the original height is disputed (there being a debate between those who argue that the original was built with a flat observatory-type apex and those who argue the original pyramid had a pyramidal apex).

This correlation between the size of the globe and the dimensions of the pyramid have staggering implications, among them the obvious conclusion that the ancients knew of the size and shape of our planet many thousands of years before conventional theories would allow. It also reinforces the conclusion that technical knowledge has not followed the conventional progression from "primitive man" to modern levels of achievement, but rather that the earliest civilizations show signs of knowledge that was later lost for centuries of darkness (theories that the earth was a sphere being considered fanciful in Europe until just over five hundred years ago).

Of course, only slightly less staggering is the fact that the designers of this earth-model selected a model scale which proves their familiarity with the subtle astronomical concept of precession. Further, all this knowledge must have been resident for some time before the pyramid itself was conceived, designed, and ultimately built.

The incorporation of a significant precessional number in the scale of the pyramid's relationship to the globe also serves to defuse the arguments of those who would argue that the 1:43,200 relationship between the pyramid's perimeter and the circumference of the globe is merely coincidental and not intended by the designers.

Elsewhere in his writings, Martin Doutré provides evidence for an ancient measurement called the reed, which was 10.5 feet (or 126 inches). This distance was also related to the Greek foot, of which there were more than one type, one of which measured 12.6 inches. Doutré points out that the recently discovered Nebra sky disc, which historians date to around 1600 BC, measures 12.6 inches in diameter (one tenth of one reed). He also notes on page 162 of *Ancient Celtic New Zealand* that the Bighorn Medicine Wheel, a stone circle in Wyoming, measures approximately 80 feet in diameter, for a circumference of 251.328 feet, or nearly 252 feet (a number occurring in other archaeological monuments around the world). This circumference of nearly 252 feet equates to 3,024 inches, which divides perfectly by 126 inches (one reed) to yield 24.

Because 126 inches (a reed) is evenly divisible by 21 inches (a royal cubit or long cubit), the length of the sides of the Great Pyramid can be calibrated in reeds as well, yielding another precessional number. Using the same measurements shown above, and dividing by 126 inches, we find the lengths of the sides to be:

East: 71.989 reeds

West: 71.978 reeds

North: 71.946 reeds

South: 72.007 reeds

The implication is that the builders of the Great Pyramid laid out sides of 72 reeds each, and that even after the depredations of centuries these measurements are still apparent. Seventy-two, of course, is the number of years required for a star to move one degree due to precession. Its presence in the measurements of the Great Pyramid makes it very difficult to dismiss its appearance in the mythology of Osiris and Set, and reinforces the argument that the murder of Osiris by Set and his 72 henchmen encodes the phenomenon of precession.

Note that we are not here attempting to correlate measurements with years or predictions of events in the past or in the future supposedly predicted by the measurements of the Pyramid and its interior passageways. Such studies are interesting, but again it's important to emphasize that the scope of this examination is to demonstrate that the dimensions of the Pyramid appear to encode precessional numbers that were known in antiquity, and therefore to reinforce the conclusions we have found in the evidence from world mythology with evidence from physical archaeology.

Further, it's important to point out that the Pyramid's incredible precision of alignment, both with the earth's cardinal directions and with the stars, strikes a telling blow against theories of continental drift. It now remains to examine other ancient structures from around the world for evidence of precise alignments that have not been abolished by long centuries of "continental drift," as they would have been if the tectonic theory were correct. If we can discover similar precessional numbers in their dimensions, so much the better.

As it happens, there are so many ancient monuments that fit both of these criteria that it is difficult to know where to begin. In his *Ancient Celtic New Zealand*, Martin Doutré thoroughly explores the incredible degree of astronomical alignment in ancient monuments, including Stonehenge and Silbury Hill in England. He also finds a plethora of precessional numbers at each. Doutré measures the distance from the center of Stonehenge to the outer face of the Heelstone, for instance, at 259.2 feet – a direct parallel to the 25,920 years for a complete circle that is obtained using one degree of precession for every 72 years (161). Similarly, he calculates that the intended distance from the base to the flat top of the huge conical mound of Silbury Hill was 259.2 feet as well.

The astronomical alignments of Stonehenge are well attested, including markings of solstitial sunrises. Even conventional archaeology dates Stonehenge's construction to 2500 BC (or at latest 2000 BC), with evidence that it may date to 3000 BC or earlier. Again, the most salient point for our argument here is that the tectonic theory is completely set on its heels by the presence of a monument built 4,500 years ago (or even 5,000 years ago) that retains precise alignments which can still be utilized today. If the continental plates are drifting by an inch a year, we would expect Stonehenge to have skidded around so much during the ensuing millennia (riding upon the Eurasian Plate that carries England with it) that its alignments would be useless to modern observers.

And it is not just the plates beneath the pyramids of Egypt or the ancient megalithic constructions of the British Isles that have left astronomical alignments intact (to the detriment of the tectonic theory). In Central America, ancient ruins contain many spectacular monuments with continuing astronomical alignments, alignments apparently undisturbed by centuries of continental drift. Some of the most famous include Teotihuacan, just north of present-day Mexico City; Chichen Itza in the Yucatan Peninsula; Uxmal, also in the Yucatan Peninsula; and Palenque, located just south and west of the Yucatan Peninsula, not far from Mexico's border with Guatemala.

In previous chapters, we presented striking visual and mythological evidence of a connection between the civilizations of ancient Egypt and the civilization or civilizations responsible for the incredible archaeological sites of Central America. Recall the similarity, first pointed out by de Santillana and von Dechend, between the scene referred to as "Uniting Egypt or Turning the Drill" (in which Horus on the viewer's left and Set on the viewer's right are engaged in pulling ropes or reeds wrapped around a central pillar or column) and the scene in the Mayan Codex Tro-Cortesianus, in which beings with more than a passing resemblance to Horus and Seth are engaged in a similar activity, using a serpent as a rope (with clear connections to similar scenes from Hindu cosmography).

The implications are staggering, and become more so the longer one considers the possibility that there were cultural connections between these ancient cultures, either directly or through some common predecessor civilization that influenced both. This possibility is vehemently rejected by conventional academia.

Any suggestion that the ancient Central and South American civilizations did not spring into being in complete isolation is treated as a horrible affront to the sensibilities of the indigenous peoples of those regions. Emotionally charged accusations of racism or ethnocentrism, however, do not address the evidence, and should have no real bearing on a quest for the truth. Again, consider the incredible similarities between the symbolism of the Seth and Horus scene and the scene in the Mayan codex, and ask if such similarities arose independently. When we consider the fact that both civilizations also built numerous astronomically-aligned pyramids and temples, and encoded precessional numbers into the dimensions of those ancient buildings, the argument that the similarities are just coincidental and that all possibility of contact must be rejected becomes more difficult to understand.

Teotihuacan is a magnificent ancient complex of pyramids and temples, with complex features whose purpose is still disputed by scholars of the site. The center of Aztec civilization when the Spaniards arrived, it was already ancient at that time. The Aztecs themselves did not claim it as their own but attributed it to an earlier people, whom they called the Toltecs, or "builders."

The complex, which contains over two and half thousand surrounding apartments, is centered around the enormous monuments known today as the Citadel (containing the Pyramid of Quetzlcoatl), the Pyramid of the Sun, the Pyramid of the Moon, and the wide avenue called the Way of the Dead (on the assumption that the magnificent pyramid complex must have been a necropolis – an assumption which is most likely incorrect, just as it is probably incorrect at the Giza complex in Egypt).

Peter Tompkins (1919 – 2007), in his 1976 work *Mysteries of the Mexican Pyramids*, highlights the astronomical alignments that have been discovered at Teotihuacan. At the center of his discussion of the modern discoveries of the complexities built into the site by the ancients is the figure of Hugh Harleston, Jr., an American engineer who spent decades at Teotihuacan attempting to unravel the meaning of the layout and proportions of the site.

Beginning in 1972, after living near the site for twenty-five years and wrestling with the challenge of unlocking the measurement used by the builders, Harleston began plotting the ratios between the largest measurements, reasoning (as Tompkins describes it) "that a comparison of *proportions* would show significant relationships despite any errors made in the reconstruction, especially if taken over large areas" (242).

He began finding recurring ratios, and eventually arrived at a measure of 1.059463 meters which, when applied to measurements of the various buildings, platforms and courtyards of the site -- as well as to the distances to certain walls that run perpendicular to the Way of the Dead at important intervals -- yielded multiples in round numbers such as 18, 24, 54, 72, 108, 144, 162, 216, and 378 (Tompkins 245). Harleston named this distance a Standard Teotihuacan Unit (STU) or a *hunab* after the Mayan word for a "measure," seen in the title of the Maya god Hunab Ku, or "giver of measures" (Tompkins 282).

Many of Harleston's original sketches are reproduced by permission in Tompkins' *Mysteries of the Mexican Pyramids*, replete with Harleston's measurements showing round numbers of *hunabs* criss-crossing the entire site. Intriguingly, many of the distances between the most significant monuments of Teotihuacan yield precessional numbers, although Tomkins does not appear to point out their precessional significance in his text. Some of these precessional numbers are highlighted in the simplified Teotihuacan diagram below.

For instance, Harleston found the distance between the Pyramid of the Sun and the Pyramid of the Moon to measure 720 *hunabs* along the axis of the Way of the Dead. The figure 72 turns up elsewhere, in the measurements of the Pyramid of the Moon itself, as well as in the plaza of the Citadel (which encloses the Pyramid of Quetzlcoatl, sometimes called the "Pyramid of the Feathered Serpent").

Multiples of 108 are also abundant, with measurements of 108 *hunabs* showing up in the courtyard of the Citadel and between certain of the walls set perpendicular to the Way of the Dead, and measurements of 216 and 324 (twice and thrice 108) appearing in the measurements of the base of the Pyramid of the Sun, the distance of the Pyramid of the Moon from the northernmost row of buildings, and in various aspects of the Citadel and the walls of the Way of the Dead.

Tomkins initiates a fascinating discussion at this point. "If Harleston's *hunab* is divided into 60 fingers, a unit of .01765 meters is obtained, which is virtually the Aztec finger, as reported by V.M.

Castillo in his *Unidaded Nohuas de Medida*, forty of which make the Aztec *bema* of .706 meter. When applied to both the Teotihuacan and Palenque complexes, Aztec fingers, palms and *bemas* give even more significant results than Harleston's *hunab*" (256). Using this *bema* measurement (which is two-thirds of Harleston's unit, or 40 fingers versus a *hunab* of 60 fingers), the perimeter of the Sun Pyramid becomes 12,960 *bemas*, and its base length 1,296 fingers (each finger being a fortieth of a *bema*) (256). While Tomkins does not take the final precessional step here, this number 12,960 is exactly one half of the precessional 25,920

Just as at Giza, numerous astronomical alignments are clearly present at Teotihuacan. Despite the site's great age, these alignments are also undisturbed by what adherents of the tectonic theory, who believe that three major plates collide along the Central American isthmus near southern Mexico, must admit

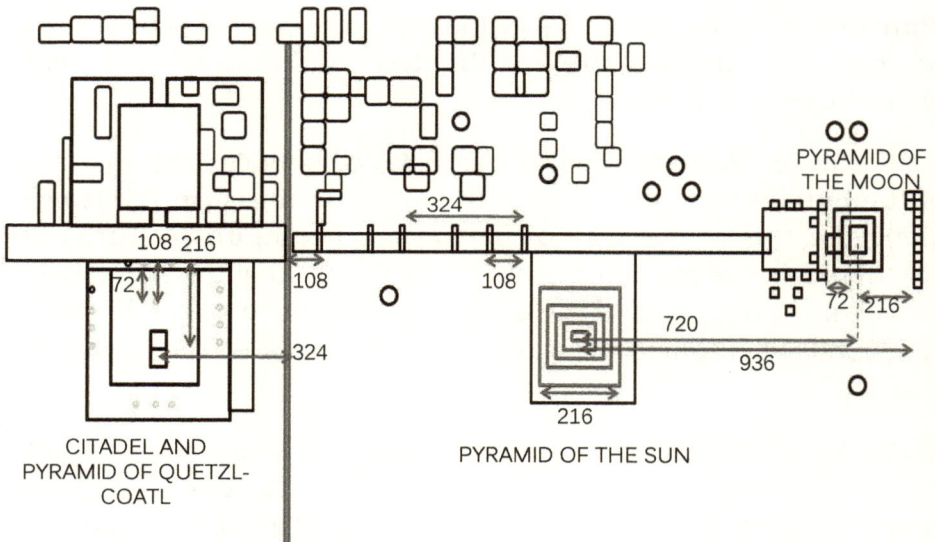

Simplified diagram of the Teotihuacan complex, showing measurements in STUs or hunabs discovered by Hugh Harleston, Jr. Beginning in 1972, after years of on-site observation, Harleston discovered that measurements of 1.059463 meters would produce whole number measurements throughout the site. In the diagram above, distances containing multiples of various precessional numbers are emphasized. Note the recurrence of 72, as well as 720 (10 x 72) and even 936 (13 x 72). Even more prevalent are 108 and multiples of 108, such as 216 and 324.

should have been many centuries of drift.

Some of these astronomical alignments have been pointed out by Graham Hancock in his *Fingerprints of the Gods*, citing Tomkins as well as other sources and his own visit. Chief among the alignments Hancock describes is the shadowplay on the western face of the Pyramid of the Sun at Teotihuacan, which occurs on the two equinoxes and creates a straight-line shadow that lasts for only 66.6 seconds (Hancock 176).

Tomkins relates the full story of Hugh Harleston's discovery of this shadow effect, when the American engineer noted the angle of the facing of the lower fourth level of the Sun Pyramid: "Noting that the lower portion of the fourth level was slightly convex and that it formed a triangle with an angle of almost 19.69 degrees to the vertical, it struck Harleston that this was also the latitude of Teotihuacan. This meant that when the sun crossed the pyramid at equinox its rays would fall onto the north face of the fourth body at the same angle of 19.69 degrees to the vertical" (251).

Harleston drew diagrams illustrating what he observed at mid-day on the equinox – at 12:35 and thirty seconds, local time, the only two shadows on the pyramid form a straight line across the west and north faces (the west side having been in shadow during the morning, but then becoming illuminated as the sun rises and moves across the sky along the ecliptic path to the south of the pyramid's location). Thirty seconds later, the shadow on the western face has begun to "erase" from south to north, leaving only the shadow along the northern face. By 12:36 and thirty-seven seconds (in other words, sixty-six seconds later), the shadow has disappeared completely, leaving only the shadow line along the northern face. Another interesting aspect of this phenomenon not added by Hancock is the fact that the section of the Sun Pyramid across which this straight shadow appears on the equinox was determined by Harleston to be . . . 72 *hunabs* in width (Tomkins 252). The henchmen of Seth, it appears, strike again – this time on the other side of the vast Atlantic.

Other important astronomical alignments at Teotihuacan detailed by Tomkins include the location of the rising sun on the equinoxes, viewed from a point along the Way of the Dead which contains a pecked cross (among many in and around the site). To an observer at the site of this particular cross, the rising sun on the equinox morning appears from the upper shoulder of the pyramid where there is a notch for one of the pyramid's levels (these levels have now been significantly altered by modern reinterpretation wrought on the pyramid in the early 1900s). This pecked cross, which has an atypical triple-foliate pattern (see diagrams), features a series of dots in its northeastern quadrant corresponding to an azimuth of 65°, an azimuth which points directly to the July solstice sun's rising point against the Sun Pyramid (Tomkins 316 – 318).

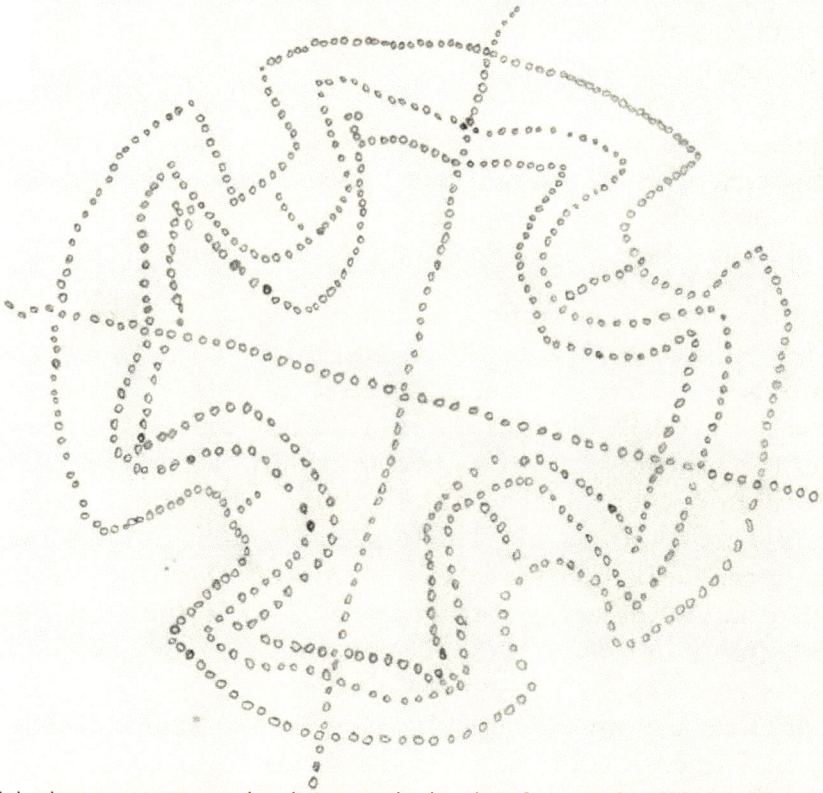

This important pecked cross is in the form of a triple cross and is somewhat more involved than the typical pecked cross pattern of a "quartered circle." This cross, with a triple foliate pattern, was discovered on March 9, 1974 by Alfonso Morales, and is located across the Way of the Dead from another pecked cross in the part of the complex known as the Viking Group. It is described in Tomkins and in a text by Dr. E.C. Krupp, *Echoes of the Ancient Skies*, on page 290.

Interestingly, important points of this triple cross appear to indicate azimuths of 72° and 108° as shown in the diagram below. These are obviously important precessional numbers. To date, I know of no analysis which points out the precessional numbers encoded in this distinctive triple pecked cross, nor of any analysis which points out the precessional significance of many of the measurements discovered by Harleston in his *hunab* postulate.

Other pecked cross markers have been located in the mountains around Teotihuacan, at elevations from which ancient observers of the heavenly phenomena could mark the point of sunrise on significant stations of its annual cycle. From one such pecked

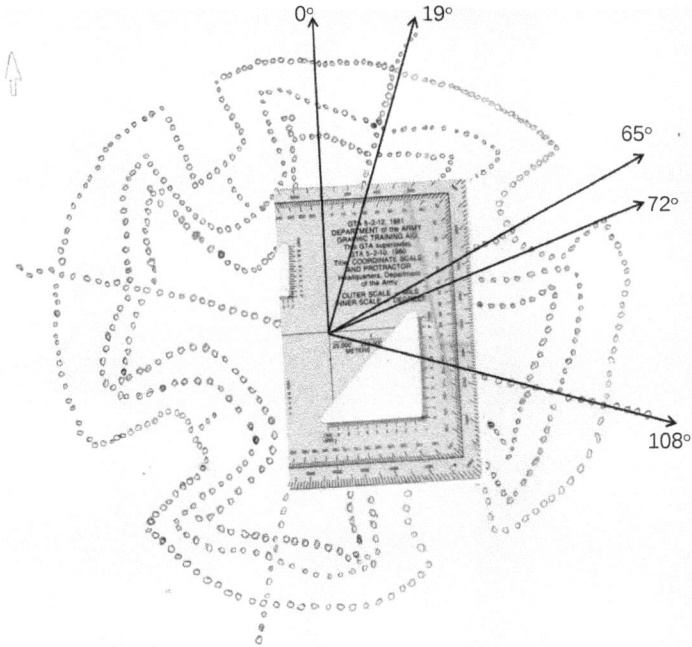

The same triple foliate pecked cross discovered on the Way of the Dead, showing the position of ture north as 0° and azimuths measured as angles from that azimuth. The direction to the sun's rise on the morning of summer solstice is at an azimuth of 65°. Note that the northern "horn" of the eastern bar of the cross delineates a 72° azimuth, and the axis through the center of this same eastern bar is alinged to a 108° azimuth. Both 72 and 108, of course, are important precessional numbers.

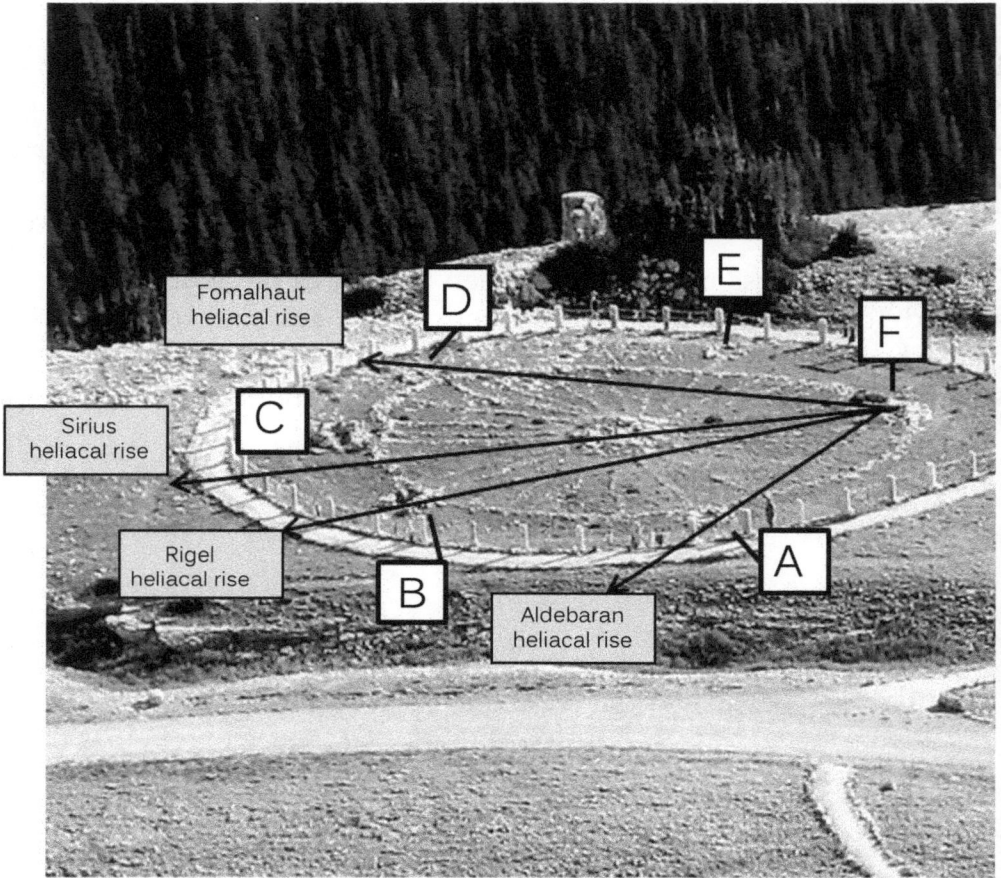

The Bighorn Medicine Wheel in Wyoming (US). Astronomer Jack Eddy in 1974 discovered important astronomical alignments, and astronomer Jack Robinson later added others. Dr. Eddy's diagram is reproduced on Stanford University's SOLAR Center website. Using the labels assigned by Dr. Eddy to the cairns at the site and a Forest Service image of the site in the public domain, these alignments are shown from the various cairns. The cairns on the outer ring are labeled A through F. The center hub, which may originally had a pole, is labeled O (not labeled here in order to reduce clutter). From cairn F, sightlines through cairns A, B, and D point to the once-yearly heliacal first rising of the stars Aldebaran (in Taurus, associated with Set by the ancient Egyptians), Rigel (in Orion), and Fomalhaut (in the Southern Fish, just below Aquarius in the northern hemisphere and not far from Capricorn; Fomalhaut is a bright 1.16 magnitude star, one of the twenty brightest in the sky). A sightline from cairn F through the center hub points to the annual first heliacal rise of Sirius, associated by the ancient Egyptians with Isis,

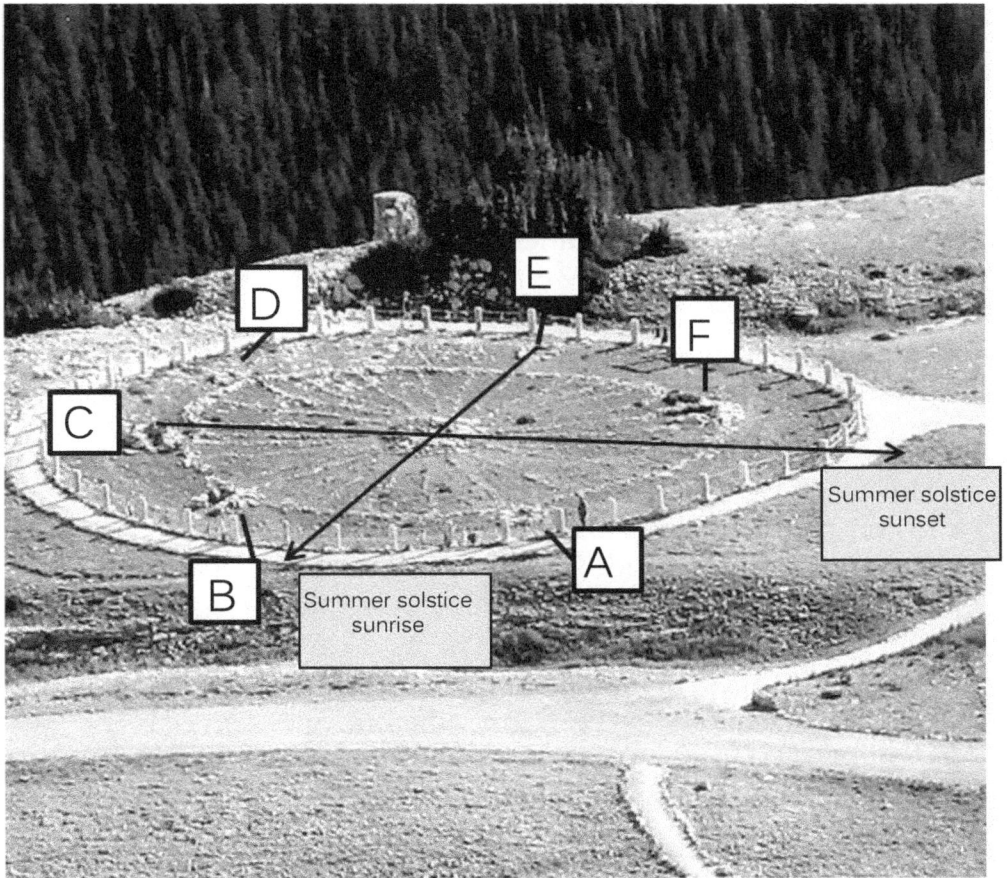

In addition to the stellar heliacal risings indicated by the sightlines depicted in the previous diagram, the Bighorn Medicine Wheel also preserves sight-lines for the rising and setting points of the sun on the summer solstice. From the cairn marked E through the central hub (which is designated O but not marked on this diagram), the sun's rising point on the July solstice is indicated, and a sighting from the cairn marked C through the central hub aligns with the sun's setting point on the same day.

cross marker on Chiconautla west of the Teotihuacan complex, discovered by Hugh Harleston and Manuel Gaitan in 1974, the azimuth to the summit of the Pyramid of the Sun is exactly 72° (Tomkins 318). In 1963, archaeologist James Bennyhoff discovered a marker at Colorado Chico on Malinalli (west of the site and north of and much closer than the Chiconautla marker) with a pecked cross with double circle, oriented such that one axis of the cross aligns with the summer solstice sunrise and winter solstice sunset points (Tomkins 319). Other modern students of Teotihuacan have discovered possible alignments with other celestial phenomena, including rising points of the Pleiades and the moon's position at the maximum northward arc that it travels during its nineteen year Metonic cycle.

This exercise of finding astronomical alignments and precessional measurements can be repeated at literally hundreds of sites around the world. We have already seen that the Bighorn Medicine Wheel in Wyoming (discussed near the beginning of this chapter) has a circumference of 251.328 feet. This number equates to 76.6047744 meters. Applying the measurements of V.M. Castillo discussed in Tomkins, wherein one Aztec finger (one-sixtieth of Harleston's *hunab*) is equal to 0.706 meters, we find that the circumference of the Bighorn Medicine Wheel is equivalent to 108.505346 of these measurements – a precessional number of great importance and unlikely to be a coincidence. I have not seen this evidence of precessional codes discussed elsewhere in conjunction with the Bighorn Medicine Wheel.

The Bighorn site preserves numerous important astronomical alignments, discovered by astronomers Jack Eddy and Jack Robinson beginning in 1974 and depicted in the diagrams on the previous pages.

Stone circles have been found around the world, in locations as far apart as Japan and Ireland. Pyramids similar in shape to those found in Mexico have been found in China. At Xi'an in Central China there are stepped pyramids with wide flat tops reminiscent of those found at Teotihuacan and Chichen Itza. The indefatigable Martin Doutré finds that the perimeter measurement of the largest pyramid at Xi'an is 2,160 feet – an obviously important precessional number that is one twelfth of 25,920. The sides of these ancient Chinese pyramids are precisely aligned to the cardinal directions of north, south, east and west (echoing the pyramids of Giza) – another refutation of ongoing continental drift. Graham Hancock has catalogued numerous ancient megalithic structures which continue to encode equinoctial and solstitial alignments to this day, such as the mysterious ancient structures on Malta in the Mediterranean.

All of these recurring examples provide overwhelming evidence that the tectonic theory is seriously flawed. They also argue for some ancient globe-spanning capability among a people who knew about precession long before conventional academia would admit, a people who knew the rate of precession to be one degree every 72 years (a far more accurate figure than was achieved by the first re-discoverers of precession in the ancient Greek and Roman world), and a people who encoded their knowledge of this number in fractions of 25,920 in their monuments.

Those who have noticed aspects of this conventional-wisdom-defying pattern have put forth various theories to explain it, from ancient visits by aliens to catastrophic impacts or near-misses by planets or other heavenly bodies. Whatever theory is correct, it is clear that the conventional timeline of man's ancient past cannot possibly be true. The hydroplate theory appears to be the most elegant of the alternative explanations offered so far. It explains all the features of the earth that gave rise to the tectonic theory, but unlike the tectonic theory, the hydroplate theory is not thrown off by the fact that all these ancient archaeological structures continue to hold their ancient alignments, and it provides a physics based reason for a "Big Roll," possibly within human memory.

The Golden Ratio and the Monuments of Antiquity

The golden ratio does not have any direct connection with the celestial phenomenon of precession, and thus a discussion of this fascinating topic may seem a bit of a tangent to the arguments put forward so far. Nevertheless, an important aspect of the argument is the theory of some form of cultural communication in man's ancient history spanning points of the globe as far-flung as Europe, Japan, China, New Zealand, and the Americas.

The extensive evidence that precessional numbers were incorporated into important ancient astronomically-aligned structures makes it extremely difficult to maintain that all of these cultures acted independently with no contact, although the purveyors of conventional academic wisdom continue to maintain exactly that untenable position. When the presence of yet another technically sophisticated series of numbers (in addition to the sophisticated understanding of the rate of precession already demonstrated – an understanding surpassing that of the most advanced theoreticians of ancient Greece and Rome, including Ptolemy's) is found among these same far-flung monuments, the weight of the evidence should be enough to convince even the most die-hard adherents of the conventional paradigm to change their position.

The golden ratio is a mathematical relationship between two unequal parts of a whole, in which (as architect György Doczi explains in his remarkable 1981 publication *The Power of Limits: Proportional Harmonies in Nature, Art, and Architecture*) "the small part stands in the same proportion to the large part as the large part stands to the whole" (2). On any given line, Doczi explains, there is only one point that will bisect it into two unequal parts with this relationship to one another and to the whole. He then goes on to demonstrate the presence of this golden ratio (often designated in modern times by the Greek letter *phi*) in a staggering array of natural growth patterns: in daisies, sunflowers,

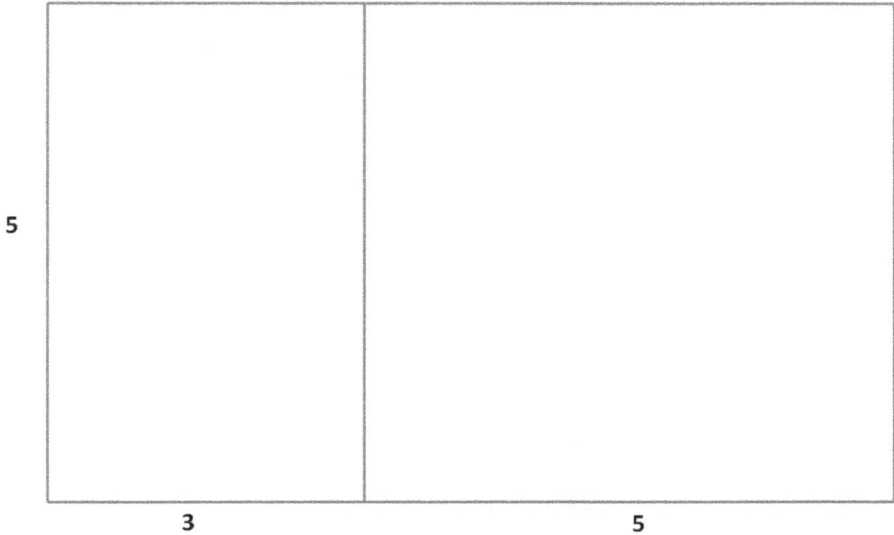

5

3   5

A Golden Rectangle. The length (3 + 5) is divided at the point of Golden Section, such that the smaller segment (3) is related to the larger segment (5) in a ratio that is equal to the ratio between the larger segment (5) and the whole (3 + 5, or 8). In this case, 5/3 is equal to 1.6667 and 8/5 is equal to 1.6. The further along the progression we go, the closer these two numbers will come to one another and to the Golden Ratio, often designated as phi by modern mathemeticians: 1.6180339887498948482045868343656381177203091 7980576 . . .
For example, if we add 8 to 5 and get 13, then 13/8 yields 1.625 (closer to the true value of phi). If we add 13 to 8 and get 21, then 21/13 yields 1.61538462. If we add 21 to 13 and get 34, then 34/21 yields 1.61904762. If we add 34 to 21 and get 55, then 55/34 yields 1.61765. Additionally, in the rectangle above, the area of the smaller section (3 x 5 = 15) relates to the area of the larger section (5 x 5 = 25) as the area of the larger section relates to the area of the entire rectangle (8 x 5 = 40). We see that 40/25 equals 1.6, and 25/15 equals 1.6667. If the rectangle instead had sides of 55 or the horizontal side (divided into smaller segment of 21 and larger of 34) and 34 on the vertical side, we would get an area for the smaller section of 714 (equal to 21 x 34) and for the larger section of 1,156 (equal to 34 x 34). Then, the ratio between the larger subarea and the smaler subarea (1,156 / 714 = 1.61904762) would approximate the ratio between the larger

apple blossoms, globe flowers, garlic buds, leaves, sea shells, clams, crabs, fishes, skates, rays, frogs, horses, peacocks, beetles, dragon-flies, butterflies, and even dinosaur bones.

The human form also exhibits golden ratio relationships, between the varying lengths of the bones of the phalanges; the distances between the top of the head, soles of the feet, and the knees; between the top of the head, soles of the feet, and nipples; between the chin, navel, and genitals; between the tip of the toe, ankle, and knee; and between lines formed by the chin, lips and nose; or nose, eyes, and brows; and many other relationships within the human body as well (Doczi, 93 – 106).

Perhaps not surprisingly, the golden ratio can be detected in art and artifacts from around the world, in all ages. Doczi points out that paper currency and credit cards typically take the form of golden rectangles (3). Temples, vases, woven baskets, Native American art, Maori tattoos (mokos), earthenware pots, geometric fabric designs, Japanese pagodas, many ancient pyramids and ziggurats, even the Parthenon of Greece and the Colosseum in Rome are shown by Doczi to contain the golden ratio at their most

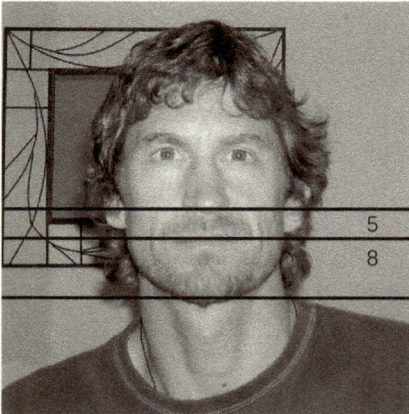

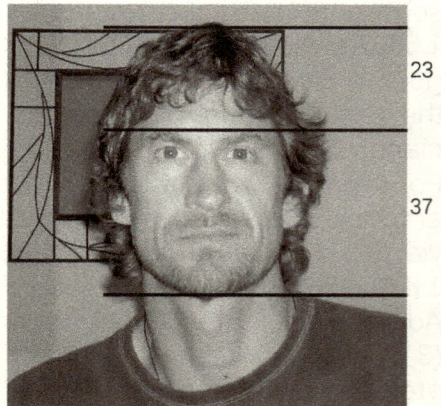

Golden Ratios are generally found in the ratio between the spacing of the chin, lips, and nose (left diagram). Also, the distance between the crown and the chin is divided at the point of Golden Section at the brows, such that the distance between the crown and brows is related to the distance between the brows and chin in the same proportion as the relationship of the distance between the chin and brows and the distance between the chin and crown (right diagram).

fundamental levels. Even the artifacts of modern technological civilization, such as the Boeing 747, are shown to contain this same phi ratio.

Doczi notes the presence of this ratio in the ancient sites Stonehenge, Teotihuacan, and Chichen Itza. He writes:

> It seems to have escaped notice so far that the architecture of Stonehenge shares the proportions of the golden section and of the Pythagorean triangle. The classical construction of the golden section applied to plan III of Stonehenge reveals that a golden relationship exists between the width of the Bluestone Trilithon Horseshoe and the diameter of the Sarsen Circle, (1:0.618 = 1.618). The same proportional construction applied to the plan of stage I shows that the rectangle formed by the Station Stones approximates the root-5 rectangle,

The Golden Ratio is further found in numerous sections of the body:
crown-sole : crown-nipples
sole-nipples : sole-knees
chin-genitals :
    genitals-navel
sole-chin : sole-genitals
crown-navel : crown-chin
crown-navel : navel-nipples
and many others.

56

34

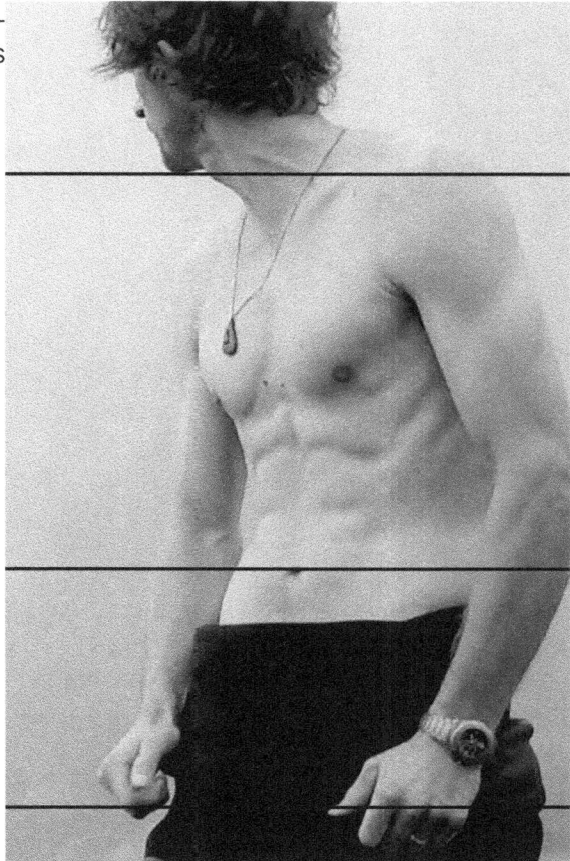

made up of two reciprocal golden rectangles. 40.

The relationships he is referring to between the width of the Trilithon Horseshoe and the Sarsen Circle are shown below on a diagram of Stonehenge. While it could be argued, given the amazing prevalence of phi ratios found in ancient and modern artifacts, that such ratios crop up unconsciously as part of the aesthetic preferences inherent in mankind, the elements of Stonehenge serve very conscious functions and were placed very deliberately, arguing that the incorporation of the golden ratio between the diameters of its various circles was not accidental or merely

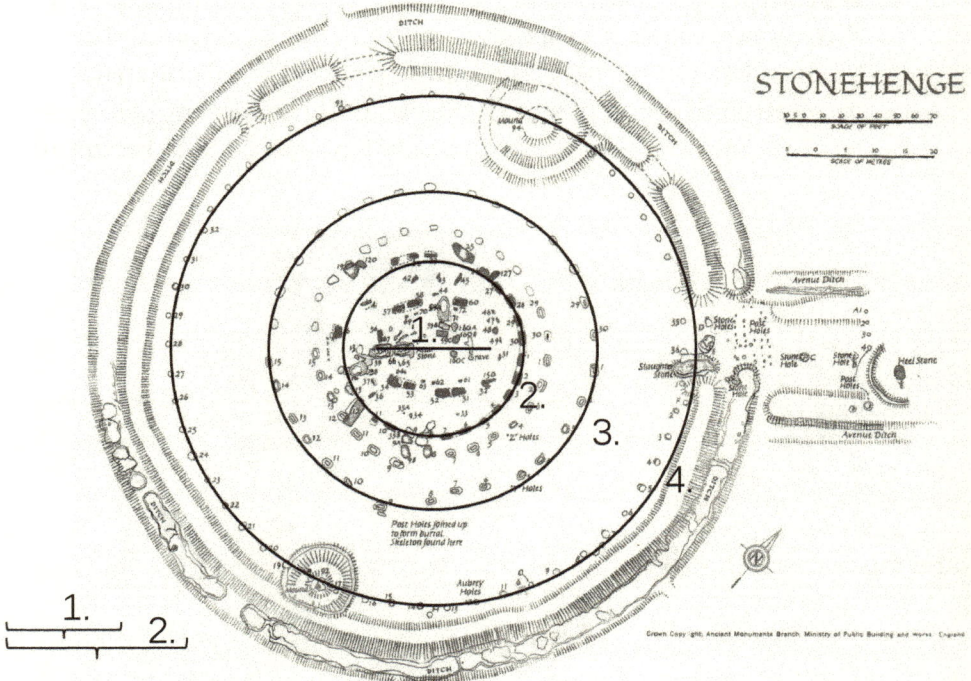

Diagram of Stonehenge showing Golden Ratio relationships. The width of the Trilithon Horseshoe shown by segment 1. is related to the diameter of the Sarsen Circle shown by circle 2. in a Golden or phi relationship, as noted by György Doczi in *The Power of Limits* (40). Martin Doutré, author of Ancient Celtic New Zealand, found numerous other phi relationships, demonstrating that the circle formed by the Y-holes (indicated by circle 3.) is a phi reduction of the circle formed by the Aubrey holes (circle 4.) The same is true of the Sarsen Circle (2.), which is a phi reduction of the Y-holes (3.)(Doutré 158-159). The length of the Trilithon Horseshoe (1.) and diameter of the Sarsen Circle (2.) are shown at the bottom left for easy comparison.

aesthetic. If we accept the proposition that Stonehenge with its various features was purpose-built for a specific functionality – and it manifestly was – then we must accept that ancient mathematicians were capable of conducting precise measurements, calculating phi, and determining the sizes they intended for the circles found in the monument based upon that mysterious golden ratio.

Martin Doutré has found numerous more instances of the golden ratio's incorporation into the details of Stonehenge. Not only is the width of the inner Trilithon Horseshoe related to the diameter of the Sarsen Circle by phi, but so also are the diameters of the Sarsen Circle and the circle formed by the "Y holes" of the monument (Doutré, 158). So also are the diameters of the "Y hole circle" and the Aubrey Circle (159). In fact, he has gone so far as to declare that "everything at Stonehenge is PHI related or connected" (167), and he backs this statement up with an incredible array of measurements and ratios involving the golden numbers.

In addition to the relationships already discussed (which are already enough to prove the proposition that the ancient builders knew the relationship we call phi, and which are diagramed in the illustration above), he finds golden rectangles (which he calls Holy Rectangles) incorporated into the site in several places, most significantly one between Aubrey Holes 43, 54, 15, and 26 and another between Aubrey Hole 40, the edge of the Slaughter Stone next to Aubrey Hole 1, 12, and 29 (Doutré, 178).

Doczi also demonstrates that the Sarsen trilithons, the famous post-and-lintel standing stones still present at Stonehenge in which two massive stones support a third stone across the top, use a 3-4-5 ratio. As shown in the diagram below, the 3-4-5 triangle is mathematically related to the golden rectangle, in that a circle drawn using the hypotenuse of the triangle as its radius will create an approximate golden rectangle with the triangle's base (from the discussion in Doczi 40-41).

Another way to visualize the relationship between a 3-4-5 triangle and phi is to imagine that the small side of 3 swings out to align with the hypotenuse of 5. The new line of length 8 will then be composed of two unequal segments of 3 and 5, such that the ratio between the subsegments (5 and 3) approximates the ratio between the whole and the larger of the two subsegments (8 and 5).

Photograph of a Sarsen trilithon showing the 3-4-5 triangle proportions.

Such a triangle is closely related to a Golden Rectangle, as shown on the next page.

The golden or phi ratio is present in numerous monuments of antiquity. Doczi demonstrates that it is present in the Great Pyramid, the Pyramid of the Sun at Teotihuacan, the Castillo at Chichen Itza, the Pyramid of the Niches at El Tajin, the Parthenon of Athens, the Temple of Athena at Priene, and in the proportions of Buddha statues from Korea and Tibet (41-47, 108-115).

Martin Doutré has demonstrated the incorporation of phi ratios in ancient stone circles from New Zealand, the cruciform meeting house which used to stand at Miringa Te Kakara in New Zealand, specific glyphs depicted by the Nazca lines (especially the Lizard, the Huarango Tree, and the Hands), the Circlestone located in Maricopa County, Arizona (in the Superstition Wilderness), the Octagon earthworks of Newark, Ohio, the moai of Easter Island,

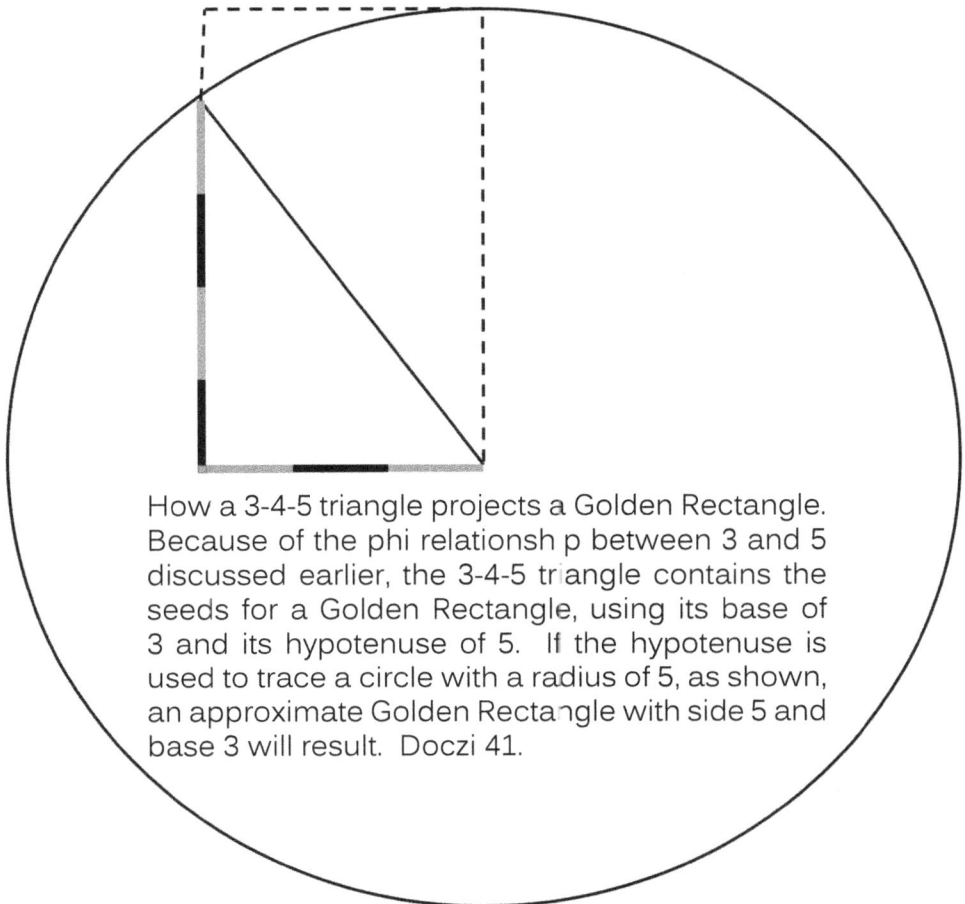

How a 3-4-5 triangle projects a Golden Rectangle. Because of the phi relationsh p between 3 and 5 discussed earlier, the 3-4-5 triangle contains the seeds for a Golden Rectangle, using its base of 3 and its hypotenuse of 5. If the hypotenuse is used to trace a circle with a radius of 5, as shown, an approximate Golden Rectangle with side 5 and base 3 will result. Doczi 41.

and many others (see his website www.celticnz.co.nz).

Doutré argues that one of the reasons for the insistent encoding of phi was the preservation of a unit of measure used by an ancient civilization, a civilization which had the capability of circumnavigating the globe and whose people eventually found themselves in all of the above locations, where their monuments all incorporated numerical patterns important to their civilization. He argues that a unit of volume essential to their economic system, the Sumerian/Babylonian Homer of 21,600 cubic inches (a precessional number) could be determined by the construction of round vessels with a diameter of the bottom measured in multiples of phi.

As we saw above in the diagram discussing the golden rectangle, the numerical value of phi is a radical number equal to 1.6180339 (rounded in this case to seven places after the decimal point). If the number 10 is divided by this number, the result is 6.18034. Doutré finds numerous ancient monuments incorporating this distance, sometimes in multiples of ten (such as 61.8 feet, the distance between the moai of Ahu North Coast IV on Easter Island, numbered 0567 and 0579 in the Rapa Nui Archaeological Database; the azimuth angle between these two moai is 129.6°, which is of course a precessional number – 12,960 being half of 25,920).

Doutré explains that the distance 61.8 feet "is in homage to a value derived from the PHI ratio, which was absolutely essential to the fabrication of precise, round, volume tubs for the market place." If vessels were constructed with base diameters of 6.18034 inches or multiples of that distance, and with sides corresponding to other coded numbers, "the final volume dispensed in the marketplace contained a code related to such things as the equatorial circumference of the earth, the cycles of the moon, the calendar count, etc. The special knowledge could never be lost because it was used on a daily basis by being encoded in the 'volume' of the cup one drank from or the tub that one bought grain from in the marketplace" (www.celticnz.co.nz/Easter Island/Easter Island 5.htm, accessed 12/14/2010).

For example, he demonstrates that a vessel with diameter of three times the above distance of 6.18034 inches and sides of height eight inches would yield a volume of 2,160 cubic inches, or 1.24 cubic feet. This volume, Doutré explains, was the original "bushel" of England, which is described in the Winchester Standard of 1696 as "any round measure with a plain and even bottom, being 18.5 inches wide throughout and 8 inches deep" (www.celticnz.co.nz/ Easter Island/Easter Island 5.htm, accessed 12/14/2010).

Doutré perceives that 2,160 is a twelfth of the entire precessional number of 25,920 and notes that the volume vessels of most of the ancient civilizations of the Mediterranean region related to this capacity: the Egyptian Theban of 11,664 cubic inches would have been 5.4 of the above-described bushels (5.4 being a tenth of 54, half of the precessional number 108); the Greek Metretes of 2,332.8 cubic inches would be 1.08 of the above-described bushels (one tenth of the precessional number 108); the Hebrew Homer, Roman Amphora, and Babylonian Archane are similarly shown to be related to the same volume (www.celticnz.co.nz/Easter Island/ Easter Island 5.htm, accessed 12/14/2010).

Amazingly, György Doczi explains in his *The Power of Limits* that the phi ratio also relates to musical wavelengths. He elaborates:

> The **1:1** proportion, which is identity, is called *unison*. The **1:2** proportion, which produces the same sound as the full string, only at a higher pitch, is called the *octave* because it reaches through all eight intervals of the scale (the eight white keys of the keyboard). The Greeks called this proportion *diapason*: *dia*, "through," *pason*, from *pas* or *pan* meaning "all." The pleasant sound of the **2:3** proportion was called *diapente* (*penta*, five), today called the *fifth*, reaching through five intervals. The consonance of the **3:4** proportion was called *diatessaron* (*tessares*, "four"), or the *fourth*.

> The 2:3 = 0.666 proportion of diapente is a close approximation of the golden section's 0.618 . . . ratio. Diatessaron is identical with the 3:4 proportion of the Pythagorean triangle. Diapason, the octave, has the 1:2 = 0.5 proportion of a rectangle composed of two equal squares, having a diagonal of √5 length, which is the joint length of

two reciprocal golden rectangles. 8.

Readers familiar with stringed instruments such as the guitar can visualize these relationships by imagining a full string which, when plucked, produces a certain note (such as a C). If the string's length is shortened by 50% (by the placement of a finger holding down the string at its exact midpoint), then plucking the remaining length will produce the same sound as the full string (another C), only at a higher pitch, as described above for the diapason or the eighth. If the string is instead held down at a point two-thirds of the way to one end, a fifth note will be produced if the new length, now 2/3 of the full length, is plucked (from C the note will now become a G). The wavelength of the sound, and the length of the string itself, are near to a phi relationship with the length of the original string and wavelength.

Readers familiar with the piano can easily visualize the same notes, which from a C go up to a G for a fifth (of course, the notes on an actual piano are striking cables which are very similar to the strings of a guitar, and the different lengths of the strings struck by the mallets within the piano create the different notes in exactly the same way as do the strings of a guitar or a violin). Note that, as Doczi later points out, the arrangement of the keyboard itself points to a golden section: from the C to the next C there are eight white keys, divided into a section of 5 and 3 by G in exactly the same way that a line of length 8 can be divided into a segment of 5 and a segment of 3. Doczi also points out that there are five black keys for every eight white keys, and the black keys are arranged in groups of 2's and 3's. "The series 2:3:5:8 is, of course, the begin-ning of the Fibonacci series, the ratios of these numbers all gravi-tating toward the irrational and perfectly reciprocal 0.618 ratio of the golden section" (10).

The Fibonacci series to which he is referring is the "summation series" in which each successive number is the sum of the two previous numbers. Beginning with 1 and 2, these can be added together to give 3, which can be added to the previous number 2 to give 5, which can be added to 3 to give 8, which can be added to 5 to give 13, and so on. This series was introduced to Europe by Leonardo of Pisa (AD 1170 – AD 1250) who brought the Hindu-Arabic numeral system to Europe and used this series as an illustration for a math problem positing the population growth of a group of rabbits in his *Liber Abaci*. While there is evidence that this series was known in India centuries earlier, it was Leonardo of Pisa (also known by his surname Fibonacci) who introduced it to the West.

The further one goes along this summation series, the more precisely the relationship between each two successive numbers will approximate phi. As we have seen, the ratio between 5 and 8 is a rough approximation of phi (0.625); if we proceed from 8 to 13 to 21 to 34 to 55 to 89 to 144 to 233, we find that the ratio between 144 and 233 is a much more accurate approximation of phi (0.618025751).

As Peter Tomkins demonstrates in his *Mysteries of the Mexican Pyramids*, the Standard Teotihuacan Unit (or hunab) discovered by Hugh Harleston and discussed in the previous chapter, works out to 1.059463 meters (the meter being a modern unit of measurement established in the end of the eighteenth century). This number is significant because a musical octave contains twelve chromatic tones, each separated in frequency by the ratio 1.059463 (or the twelfth root of two – a number which, when multiplied by itself twelve times, will produce two). A frequency of 65.41 hertz (meaning a vibration rate of 65.41 cycles per second) will produce a low C note. Increasing the cycles per second by a factor of 1.059463 will give a vibration rate of 69.30 hertz (actually 69.299 hertz), which will produce a C#. Multiplying this frequency by the factor of 1.059463 will move the vibration up another half step, to a low D note. This progression can be continued twelve times until a frequency of twice the original is reached, at which point the next higher pitch of C will be attained.

Of course, the above relationship between the hunab of Harleston measured in meters and the factor needed to move up the musical chromatic scale would appear to be simply an incredible coincidence, since the ancient builders did not use meters. However, recall that the original meter was conceived in the eighteenth century as a fraction of the earth's meridian from the equator to the north pole. Tomkins points out that Harleston's hunab could have been selected by the ancient builders of Teotihuacan as a fraction of the earth's diameter in much the same way: "Harleston wondered if the ancient builders of Teotihuacan might have arrived at such a basic unit by dividing the polar diameter of the earth not into 20 million parts as postulated by Herschel and Smyth for a unit of .6356 meters, but into 12 million parts for a unit of 1.059 meters. For neither one of these units was knowledge of the meter required [. . .]" (244).

This still would not explain the relationship between the hunab as expressed in the modern meter and the twelfth root of two (the factor by which each frequency of the chromatic scale is separated from the adjacent tone's frequency). However, if the earth's circumference in meters (about 40 million) is divided by pi to yield an approximate diameter (since a circle's circumference is equal to pi times the diameter) and the resultant number is divided into 12 million parts as postulated above by Hugh Harleston, a unit of about 1.059 meters would be achieved; if we grant such geodetic knowledge to the ancients, then it is not inconceivable that they could have also divided the circumference into 40 million parts and known of the ratio between the two measures. Such an explanation may be a stretch, but it would yield two measures related to one another as 1 is to 1.059, and thus by the same ratio as one chromatic tone's frequency is to its neighbor's.

Be that as it may, the presence of phi ratios in ancient monuments is indisputable, as is the relationship between the phi ratio and musical harmonies, particularly the fifth. The incorporation of phi into art can be explained as driven by the aesthetic pleasure such proportions produce in humans, and thus skeptics could say that ancient sculpture and even temples and pyramids incorporating phi ratios could have been created merely by artistic taste with no knowledge of the mathematics required. However, the incorporation of the same golden ratios into the diameters of stone circle rings that served a more utilitarian than artistic purpose seems to argue that the ancients were aware of phi and consciously encoded it into their creations. The relationship of phi to the ancient units of measurement argues the same conscious knowledge.

The widespread use of this fascinating and mathematically sophisticated ratio (and the precision with which it was incorporated
into widespread ancient monuments) argues again for a common
civilization influencing these widely dispersed sites, or a common
predecessor. It is unlikely that such a precise knowledge of phi,
and such an obsession with it, would have arisen over and over
again in widely dispersed and culturally isolated peoples, especially when we consider the modern history of this mathematical
concept and the degree to which relatively advanced civilizations remained ignorant of it for long centuries until coming into
contact with other cultures who had preserved its knowledge.

More evidence from archaeology:
The Cult of Mithras

In the preceding discussion, we have seen extensive evidence from the mythological artifacts of the ancient world (including the mythology of North America, Central America, South America, and the Pacific Oceania), and from the archaeological artifacts scattered across the globe (from Egypt to Mexico to China) to support the theory that mankind at a far earlier date than conventional theorists will admit possessed a sophisticated understanding of the earth we live on and its path through the heavens, and in particular the precessional wobble that brings about the subtle but inexorable march of the stellar backdrop to the solstitial and equinoctial stations of the year.

Further, we have seen connections between these widely dispersed and almost ubiquitous artifacts that suggest a common ancestry, either from some ancient culture known to us (such as the ancient Sumerians, Phoenicians, or Egyptians, for example) which was able to and did traverse the entire globe, or from some common forgotten predecessor of all these cultures which left its fingerprints on all of them.

We have examined extensive evidence from de Santillana and von Dechend's groundbreaking study of this subject, *Hamlet's Mill*, probably the first serious text since the 1800s to put forth a cohesive argument for this forgotten ancient knowledge and ancient connection. Looking at other archaeological sites, we have found further evidence beyond the massive volume of evidence offered by the authors of *Hamlet's Mill*, including the evidence from the medicine wheels of North America, the precessional numbers encoded in measurements of monuments in Egypt, China, the British Isles, North America and Central America, the precessional numbers encoded in the angles indicated by the pecked crosses found in Teotihuacan, and the presence of golden ratios used in ways that must have been conscious in ancient monuments from Stonehenge to New Zealand.

De Santillana and von Dechend demonstrate fairly conclusively that certain recurring motifs, such as the whirlpool and the tree (or the whirlpool and the vine), or the turning millstone that slips off of its hub axle or "spindle," encoded metaphorical descriptions of the vast turning of the celestial mechanism and its ages-long precession through the zodiac. Another pattern they found appearing in myths across many cultures is the coded description of fire-related motifs to represent the equinoctial colures, those imaginary "hoops" which define at their intersection with the path of the ecliptic the two important stations of the year at the equinoxes. Remember that it was the spring equinox which in antiquity was associated with the start of the new year and which determined the current age (from the Age of Taurus to the Age of Aries to the Age of Pisces and on to the Age of Aquarius). Once we are alert to the idea that fire is often a metaphor in ancient myth for the equinoctial colure and the equinoxes, we can begin to unlock several important ideas.

For example, de Santillana and von Dechend examine in detail the myth of Prometheus who brought fire to mankind, and the myth of Phaethon whose mishandling of his father's chariot of the sun scorched parts of the earth and led to his own death, and suggest that each of these "world-changing" events represent the changing of an age – the movement of the "fire" of the sun to a new station in the celestial watch-bezel, so to speak. When Prometheus brought fire to mankind, it symbolized a major change in the power structure of the celestial realm, just as the shift in the equinoxes ends one age and begins another. The myth of Phaethon similarly represents the movement of the sun caused by the phenomenon of precession, but in a more negative way: the sun has strayed off course, and the order of the heavens is upended.

De Santillana and von Dechend also note the presence of fire in the rainbow bridge to Asgard in Norse myths, which was too hot for the snow-giants and jotuns to ascend and which was guarded by Heimdal, a personage whom de Santillana and von Dechend demonstrate to have had equinoctial associations, particularly in his possession of the mighty Gjallarhorn which he blew to signal the end of the age.

They explore connections between Heimdal, who is described in the Eddas as being "the son of nine mothers," and the fire deity Agni in the *Rigveda*, as well as with Agni's son Skanda in the *Mahabharata* (157). They posit a connection to the nine goddesses who in the Norse Eddas turn the mill that grinds out the fates of men. De Santillana and von Dechend elaborate on this important mythological pattern:

> The nine grim goddesses who "once ground Amlodhi's meal," working now that "host-cruel skerry quern" beyond the edge of the world, are in their turn only the agents of a shadowy controlling power called Mundilfoeri, literally "the mover of the handle." The word *mundil*, says Rydberg, "is never used in the old Norse literature about any other object than the sweep or handle with which the movable millstone is turned," and he is backed by Vigfusson's dictionary which says that "mundil" in "Mundilfoeri" clearly refers to "the veering round or revolution in the heavens."
>
> The case is then established. But there is an ambiguity here which discloses further the depths in the idea. "'Moendull' comes from Sanskrit 'Manthati,'" says Rydberg, "it means to swing, twist, bore (from the root *manth-*, whence later Latin *mentula*), which occurs in several passages in the Rigveda. Its direct application always refers to the production of fire by friction."
>
> So it is, indeed. But Rydberg, after establishing the etymology, has not followed up the meaning. The locomotive engineers and airplane pilots of today who coined the term "joy stick" might have guessed. For the Sanskrit Pra-mantha is the male fire-stick, or churn stick, which serves to make fire. And Pramantha has turned into the Greeks' Prometheus,

a personage to whom it will be necessary to come back frequently. What seems to be deep confusion is only two differing aspects of the same complex idea. The lighting of fire at the pole is part of that idea. But the reader is not the first to be perplexed by an imagery which allows for the presence of planets at the pole, even if it were only for the purpose of kindling the "fire" which was to last for a new age of the world, that world-age which the particular "Pramantha" was destined to rule. [. . .]

It should be stated right now that *"fire" is actually a great circle reaching from the North Pole of the celestial sphere to its South Pole*, whence such strange utterances as *Rigveda* 5.13.6: "Agni! How the felly the spokes, thus you surround the gods." (Agni is the so-called "fire-god," or the personified fire.) The *Atharva Veda* says, moreover, that the fire sticks belong to the *skambha*, the world's axis, the very skambha from which the Sampo has been derived. 139-140.

In other words, the "great circle" to which de Santillana and von Dechend refer is the great equinoctial hoop or colure stretching from equinox to equinox where it intersects the ecliptic, and passing through the celestial north and south poles (see the diagrams in chapter 3, page 62). The shifting of that great circle along the ecliptic gives rise to the successive ages – which, as the authors explain, is encoded in mythology as the kindling of a new fire which is to last for a new age.

They point out that this Promethean task is found in mythologies around the world. If the connection between ancient Hindu Vedas and ancient Norse mythology does not seem a staggering enough stretch to the reader (if the conventional theories can somehow account for an ancient civilizational cross-pollenation between ancient India and Viking Scandinavia), the presence of the same motif in the mythology of the North American Indians is totally beyond explanation by the conventional theories (other than the dubious theory that such mythological motifs just sprang up completely independently of one another). We have already cited the Catlo'ltq (native to British Columbia in the Pacific Northwest of Canada) myth of the stealing of fire by the stag, which hid embers in its coat to bring them to mankind (discussed in Hamlet's Mill, 318-319). To obtain the sacred fire, the deer had to leap through a door which clanged open and shut (just like the Symplegades or clashing rocks in Greek mythology of Jason and the Argonauts, which we previously learned was a metaphor for the shifting equinoxes; the Golden Fleece that Jason and the Argonauts pursued being yet another metaphor for the sacred fire).

Further confirmation of the interpretation offered by de Santillana and von Dechend of fire as a symbol for the equinoctial colure which divides the ages is found in David Ulansey's groundbreaking examination of the plentiful archaeological remnants of the cult of Mithras which flourished in the Roman Empire concurrent to the dawn of Christianity. In *Origins of the Mithraic Mysteries: Cosmology and Salvation in the Ancient World* (Oxford UP, 1989), Ulansey outlined a compelling argument that the conventional attempts to explain Mithraism were flawed, and that they could be better understood in reference to the mechanisms of precession.

The cult of Mithras became extremely widespread during the Roman Empire and was one of the most important "mystery religions," (another important mystery religion was the cult of Isis). The cult of Mithras was also the chief rival to Christianity during the first centuries of its existence. Because it was a mystery religion, in which the teachings and rites were only revealed to initiates, the practitioners themselves left us almost nothing in writing about their beliefs, and modern scholars have had to piece together theories based for the most part upon the extensive archaeological remains of the Mithraic monuments, underground temples known as *mithraea*.

The first mention of the cult is by Plutarch (AD 46 – some time after AD 120), whom we have already met as the author of the first comprehensive discussion of the Isis and Osiris story in existence today, writing in his *Lives*, specifically in his *Life of Pompey*. From the 1917 translation by Bernadotte Perrin in volume V of the Loeb Classical Library edition, we read Plutarch's description of the pirates who came to dominate the Mediterranean to such a degree that Rome called upon Pompey, already a celebrated and beloved general, to subdue them. This took place in 67 BC. We will examine Plutarch's description of the pirates in detail, because the aspects of these pirates, who apparently first spread the Mithraic mysteries, is important to Ulansey's theory:

> The power of the pirates had its seat in Cilicia at first, and at the outset it was venturesome and elusive; but it took on confidence and boldness during the Mithridatic war, because it lent itself to the king's service. Then, while the Romans were embroiled in civil wars at the gates of Rome, the sea was left unguarded, and gradually drew and enticed them on until they no longer attacked navigators only, but also laid waste islands and maritime cities. And presently men whose wealth gave them power, and those whose lineage was illustrious, and those who laid claim to superior intelligence, began to embark on piratical craft and share their enterprises, feeling that the occupation brought them a certain reputation and distinction. There were also fortified roadsteads and signal-stations for piratical craft in many places, and fleets put in here which were not merely furnished for their peculiar work with sturdy crews, skilful pilots, and light and speedy ships; nay, more annoying than the fear which they inspired was the odious extravagance of their equipment, with their gilded sails, and purple awnings, and silvered oars, as if they rioted in their iniquity and plumed themselves upon it. Their flutes and stringed instruments and drinking bouts along every coast, their seizures of persons in high command, and their ransoming of captured cities, were a disgrace to the Roman supremacy. For, you see, the ships of the pirates numbered more than a thousand, and the cities captured by them four hundred. Besides, they attacked and plundered places of refuge and sanctuaries hitherto inviolate, such as those of Claros, Didyma, and Samothrace; the temple of Chthonian earth at Hermione; that of Asclepius in Epidaurus; those of Poseidon at the Isthmus, at Taenarum, and at Calauria; those of Apollo at Actium and Leucas; and those of Hera at Samos, at Argos, and at Lacinium. They also offered strange sacrifices of their own at Olympus, and celebrated there certain secret rites, among which those of Mithras continue to the present time, having been first instituted by them.

But they heaped most insults upon the Romans, even going up from the sea along their roads and plundering there, and sacking the neighbouring villas. Once, too, they seized two praetors, Sextilius and Bellinus, in their purple-edged robes, and carried them away, together with their attendants and lictors. They also captured a daughter of Antonius, a man who had celebrated a triumph, as she was going into the country, and exacted a large ransom for her. But their crowning insolence was this. Whenever a captive cried out that he was a Roman and gave his name, they would pretend to be frightened out of their senses, and would smite their thighs, and fall down before him entreating him to pardon them; and he would be convinced of their sincerity, seeing them so humbly suppliant. Then some would put Roman boots on his feet, and others would throw a toga round him, in order, forsooth, that there might be no mistake about him again. And after thus mocking the man for a long time and getting their fill of amusement from him, at last they would let down a ladder in mid ocean and bid him disembark and go on his way rejoicing; and if he did not wish to go, they would push him overboard themselves and drown him.

This power extended its operations over the whole of our Mediterranean Sea, making it unnavigable and closed to all commerce. This was what most of all inclined the Romans, who were hard put to it to get provisions and expected a great scarcity, to send out Pompey with a commission to take the sea away from the pirates. 175-179.

This first existing description of the secret rites of Mithras, which Plutarch notes continued to the present time during which he wrote, identifies the pirates of Cilicia as being those who "first instituted" these rites. Cilicia at the time was a Roman province located in the area historically known by that appellation and mentioned in Homer, situated along the Anatolian coastline bordered on the north by the Taurus Mountains and reaching inland to the Amanus Mountains. It contains the broad plain of Issus and important ancient cities including Salamis, Adana, and Tarsus.

Ulansey notes that Plutarch's explanation and ascription of the origin of the cult to the pirates active at the time of Pompey (in the first century BC) is generally "accepted as reliable" and cites later evidence of Mithraism in Tarsus, the capital of the province of Cilicia, as corroborating Plutarch's assertion (40-41). He notes that scholars of Mithraism such as Franz Cumont (1868 – 1947) and his student and prodigious scholar Maarten J. Vermaseren (who published the definitive corpus of Mithraic monuments, the Corpus Inscriptionum et Monumentorum Religionis Mithriacae, or CIMRM) both accepted Plutarch's account as probable.

Ulansey explains that:

> owing to the obscurity of Mithraic iconography and the general absence of any ancient explanations of its meaning, the *internal* aspects of Mithraism (i.e., the beliefs and teachings of the cult) have resisted the attempts of scholars to decipher their secrets. But the Mithraists did leave to posterity a key for unlocking the inner mysteries of their religion. For although the iconography of the cult varies a great deal from temple to temple, there is one element of the cult's iconography which was present in essentially the same form in every mithraeum and which, moreover, was clearly of the utmost importance to the cult's ideology: namely, the so-called tauroctony, or bull-slaying scene, in which the god Mithras, accompanied by a series of other figures, is depicted in the act of killing a bull. This scene was always located in the central cult-niche of the mithraeum. The fact that this iconographically fixed representation appeared in the most important place in every mithraeum forces us to conclude that it was of central importance to the cult's ideology and that its meaning, if we can decipher it, holds the key to the mystery of Mithraism. 6.

Ulansey recounts that Belgian scholar Cumont's explanation of the tauroctony scene dominated Mithraic scholarship from the late 1800s through the 1970s. Cumont advanced the argument that Mithraism came to the Roman Empire through Persia, which accords with recorded Roman belief that Persia was the origin of the cult of Mithras (the cult was often referred to as the "Persian

Tauroctony from the mithraeum at Nersae (Italy, modern Nesce) showing scorpion, snake, dog and raven in their customary locations (the snake's location may change from one tauroctony to the next). Ulansey notes that the lion and the cup are not always present and that when they are, they are usually present together. From this he argues that they represent the solstices, which were in Leo and Aquarius when the equinoxes were in Taurus and Scorpio (and that the cup thus represents Aquarius rather than the constellation Crater).

Mysteries" in ancient Rome). Cumont theorized that Mithras was etymologically descended from the Persian deity Mithra; however, a major problem with the theory, Ulansey explains, is that nowhere in ancient Persian myth is Mithra involved in slaying a bull. Cumont co-opted a Zoroastrian legend (not recorded until the 9th century AD, but believed to incorporate earlier Persian tradition) in which a figure not associated with Mithra – in fact, the embodiment of cosmic evil – slays a bull, and argued that this was the antecedent for the bull-slaying scene in Mithraism. Because in that legend the various life forms spring forth from the slain bull, Cumont believed that the various animals depicted in the tauroctony scene – a dog, a snake, a raven, a scorpion, a lion –

represent new life being born from the sacrifice of the bull.

In 1971, Mithraic scholars began to question the validity of Cumont's analysis, particularly with the publication of two papers critical of his methods and conclusions at the First International Congress of Mithraic Studies at Manchester University, one by R. L. Gordon and one by conference organizer John Hinnells. Both authors found several serious problems with the connection to Iranian mythology and concluded that Cumont's hypothesis was flawed. Subsequently, scholars rediscovered the hypothesis of German scholar Karl Bernhard Stark (1824 – 1879), who had argued for an astrological origination for the symbols in the tauroctony, a hypothesis that Cumont had severely criticized and which had thus been neglected for over seventy years.

Ulansey takes this line of inquiry and argues that the constellations depicted are those which were found along the celestial equator when it is below the ecliptic, saying: "The hypothesis which I would like to put forward here is that the tauroctony does indeed represent the celestial equator, but that it represents the

Cautes (left, torch pointing upwards) and Cautopates (right, torch pointing downwards) in the same tauroctony as previous diagram. This tauroctony (a bas-relief in marble) is dated to AD 172.

celestial equator *as it was when the equinoxes were in Taurus and Scorpius* rather than in Aries and Libra" (51). Now, to understand what he means, we must think back to the discussion in chapter 3, and recall that in winter in the northern hemisphere, the ecliptic is above the celestial equator at night (below it during the day). In summer, the opposite situation obtains, and the ecliptic is above the celestial equator during the day and below it at night. I believe Ulansey is correct, and that the tauroctony depicts the winter celestial equator, above the ecliptic, from fall equinox to spring equinox. The time between the equinoxes could be either from spring to fall or from fall to spring, but it appears to be fall to spring, an interpretation consistent with the celebration of the birthdate of Mithras near the winter solstice, right on or around December 25. Ulansey demonstrates that the constellations that would have been crossed by the celestial equator when the equinoxes were in the Scorpion and the Bull were: Taurus the Bull, Canis Minor the Little Dog, Hydra the Serpent, Crater the Cup, Corvus the Raven, and Scorpio the Scorpion (51).

Ulansey argued that the case for identification of these figures with constellations is sealed by the presence of two distinctive figures found flanking the scene in most of the tauroctonies, the figures known as Cautes and Cautopates (names we know due to dedicatory inscriptions extant in mithraea), who bear torches, one pointing upwards (Cautes) and one pointing downwards (Cautopates), and who are usually dressed (like Mithras) in tunics and Phrygian caps, and whose legs are usually crossed.

Ulansey argues that these figures represent the equinoxes themselves, when the path of the sun in its journey towards either solstice crosses the celestial equator, either crossing above at the spring equinox (when the daylight hours become longer than the hours of darkness) or crossing below at the fall equinox (when the daylight hours become shorter enroute to the winter solstice or shortest day of the year in the northern hemisphere).

He bolsters this argument by the position of the torches (pointing upwards to represent the sun crossing north of the ecliptic during the day or downwards to represent the sun crossing south of the ecliptic during the day), as well as attendant symbolism sometimes found accompanying these figures, including a tree in flower (for spring, with Cautes) and in fruit (for fall, with Cautopates). Most significantly, there are several instances in which Cautes and Cautopates are represented with symbols for a bull and a scorpion, including an instance in which each of the figures carries an actual bull's head or scorpion's body in the crook of his elbow. Ulansey argues that this clearly identifies the pair with the equinoxes, with Cautes (spring, torch pointing upwards) carrying the symbol of Taurus (representing the spring equinox in Taurus) and Cautopates (fall, torch pointing downwards) carrying the symbol of Scorpio (representing the fall equinox in Scorpio).

The central thrust of Ulansey's thesis is the proposition that the new religion of Mithraism, which surfaced around Cilicia in the first century BC based on the account of Plutarch, was a response to the new discovery of the phenomenon of precession by Hipparchus. Ulansey speculates (his own word) that the discovery that the staid mechanism of the cosmos, which seems to wheel on through the millennia with mechanical regularity around an Axis that (in the words of the poet Aratus, whom we have discussed before) "shifts not a whit," was in fact being moved inexorably by some force even greater than the zodiac. Ulansey argues that this discovery would have been world-shattering, particularly to the Stoics who were prevalent in Tarsus around the same time (79-81).

Ulansey cites ancient authors, including Cicero, who attested to the Stoics' well-known belief in an irresistible fate which ordained all that would come to pass, a fate to which they must submit without complaining or emotional distress. This belief was intimately connected to their keen interest in astrology, and the belief that the stars ruled the fate of men. Further, the Stoics are described by many ancient sources, including Cicero, as believing that the present world age would come to a cataclysmic end – "consumed in flame," Cicero tells us, and that "from this divine fire a new universe would be born and rise again in splendor" (Ulansey 74, citing Cicero's description of the Stoic belief in *Nature of the Gods*).

From this historical fact, Ulansey posits that the revelation by Hipparchus of the precession of the equinoxes would have been perceived to have world-shaking impact, sufficient to create a new deity capable of moving the established order of the heavens. He theorizes that the origin of the Mithraic mysteries lay in this new revelation of heavenly motion, and that the tauroctony encapsulates the mystery of the new knowledge: by showing the death of the bull, it was encoding the ending of the previous cosmic age (the Age of Taurus, from around 4,000 BC to 2,000 BC), a clear marker of the newly understood phenomenon. The further identification of Cautes (spring equinox during the age previous) with Taurus and Cautopates (fall equinox during the age previous) with Scorpio provides confirming evidence for aspects of Ulansey's theory.

Ulansey found further support for his theory of the genesis of the religion in the fact that the pirates of Cilicia were described by Plutarch as drawing even members of the intellectual class, lending credence to the possibility that Stoics of Tarsus found their way to a life of piracy, and that the incipient religion thus began to overspread the empire (88). He points out that sailors are already predisposed to knowledge of and reverence for the stars, making them "particularly receptive to religious teachings involving a deity whose essential characteristic was his power over the stars" (88).

Ulansey's thesis was strongly criticized by (among others) Professor Noel M. Swerdlow of the University of Chicago, in *Classical Philology* volume 86, number 1 (January, 1991), in a review. The entire text of Professor Swerdlow's argument is available online at http://people.sc.fsu.edu/~dduke/lectures/swerdlow-mithras.pdf

Swerdlow dismissively states, in response to Ulansey's identification of Cautes and Cautopates with the equinoxes based on the spring and fall images and the presence of Taurus and Scorpio symbology in some of the scenes, "It need hardly be pointed out that all this necessarily shows is an identification of Cautes with Taurus and spring and Cautopates with Scorpio and autumn, without any reference to the equinoxes themselves, a point to which we shall return" (53).

Professor Swerdlow argues that because Scorpio and Taurus were no longer at the equinoxes but had become associated with the seasons of summer and winter in the age after 2,000 BC (he cites ancient almanacs to prove it), that this indicates that Cautes and Cautopates could easily represent not the specific dates of the spring and fall equinoxes but rather the seasons of summer and winter. He then states (not without some justification) that without the identification of Cautes and Cautopates with the equinoxes, "Ulansey's entire argument fails" (59).

It is true that if the torchbearers do not represent the equinoxes, then the identification of the tauroctony with the end of the Age of Taurus has serious problems, but has Swerdlow really dismantled Ulansey's argument by showing that the constellations of Taurus and Scorpio had moved to summer and winter in the ensuing two millennia before the first mention of the Persian mysteries? After all, Ulansey argued that the originators of the Mithraic cult consciously chose to incorporate symbols of the end of the previous age. Further, the tree symbology next to Cautes and Cautopates (of a tree in flower and a tree in fruit) are symbolic of Spring and Fall, and not of Summer and Winter (in which case we would expect to find a tree in leaf and a tree with bare branches).

Even more damning to Swerdlow's hasty dismissal of Ulansey's argument is the abundant evidence cited at the beginning of this chapter, and piled up extensively in de Santillana and von Dechend, showing that ancient civilizations (long before the arrival of the first description of the Persian mysteries) understood the equinoxes to be associated with fire. The authors of *Hamlet's Mill* devote extensive space to the importance of the fire symbology in the myth of Prometheus (symbolizing the first Golden Age of mankind), the myth of Phaethon (symbolizing the destruction of one age or "sun" and the conflagration of the world, which still bears the marks of the cataclysm he caused), the myth of the Argonauts who were pursuing the shining skin of a ram (the new "sun" of the Age of Aries) and who had to pass through the clashing rocks of the Symplegades as part of their quest (as we have already discussed in some detail, a myth which is significantly echoed among Native American mythology by the retrieval of fire by the stag, who also must pass through the clashing door to bring fire to the people, a clear echo of the same concept), and the Norse myth of Heimdal the guardian of the fiery rainbow bridge, which was destroyed in the world-ending conflagration by Surt, the fire-demon of Muspelheim, to usher in the present age and destroy the previous one.

Given this pedigree, it is perhaps hasty to dismiss the possibility that the torch-bearers are equinoctial symbols, as Professor Swerdlow does. It is especially ill-advised to do so given the clear equinoctial symbology of the figures' crossed legs (which Swerdlow himself mentions as potentially indicating the crossing of the celestial equator and ecliptic, even if Ulansey does not specifically make that connection), and the torch pointing upwards for the spring equinox (when the ecliptic passes above the equator and ushers in the half of the year in which day is longer than

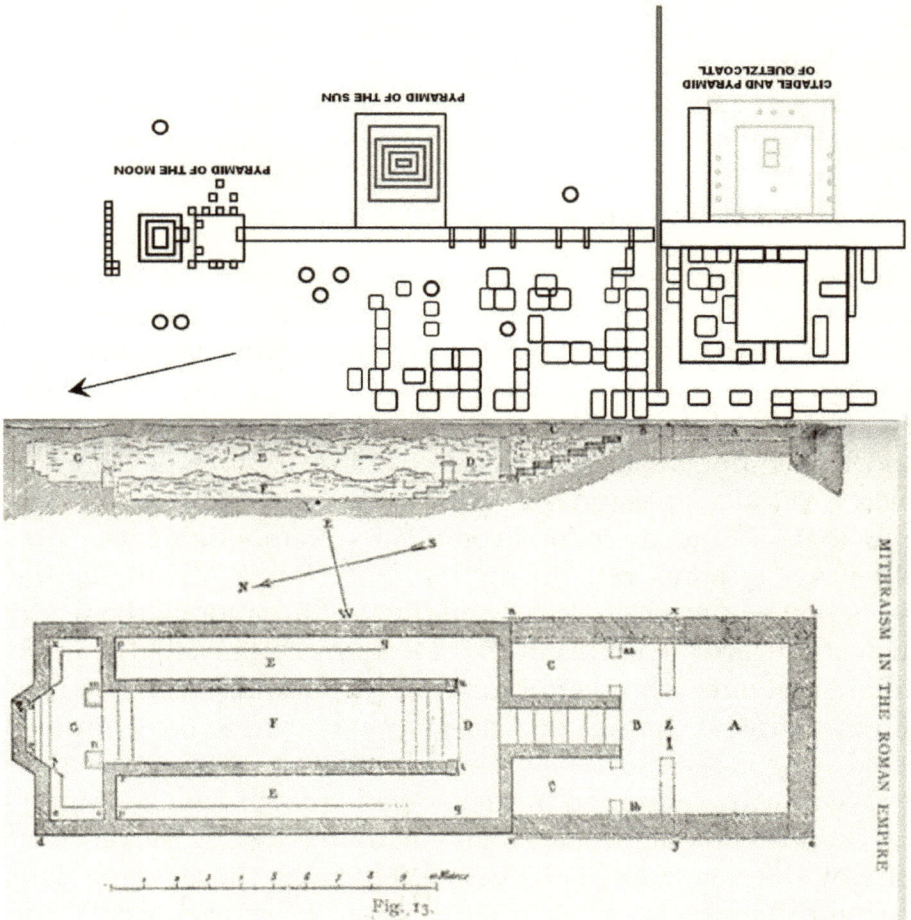

Fig. 13.

night) and downwards for the fall equinox (when the ecliptic passes below the equator and ushers in the half of the year in which night is longer than day).

On the other hand, it is quite clear from the mass of other evidence we have examined that it is difficult to accept (as Professor Swerdlow also does not) that the Mithraic symbology was a response to the recent "discovery" of precession by Hipparchus. Swerdlow points out many valid reasons why it is ill-advised to believe that Hipparchus' tentative conclusions could have led so rapidly to world-shaking religious reconsideration by the Stoics of Tarsus, who then rapidly concocted a symbology based upon the zodiacal positions dominating the heavens some two thousand years previous. Far more plausible is the likelihood that such secret astronomical coding had survived in many ancient near eastern civilizations (including the region of Anatolia) and that its manifestation in the mysteries of Mithras was yet another in a long history of incarnations which we find around the world (albeit a particularly explicit incarnation).

Further indication that the origin of Mithraic symbology had to do with a knowledge of precession more ancient than that outlined by Hipparchus is found in the dimensions of the mithraea themselves. For instance, in the surviving Mithraeum of London (on the bank of the Walbrook Stream, CIMRM 820), the length of the original temple has been measured to be 18.3 meters – or 720 inches! Although I am not aware that anyone else has pointed this out, this measurement is a clear precessional number, and one based upon a knowledge of precessional rate more accurate than any calculated by Hipparchus.

In the diagram of the Mithraeum of Heddernheim depicted on page 53 of Cumont's original text *Mysteries of Mithra*, the length of the entire temple is divided into an area from the entrance to the first wall (this section is called promenade and sacristy in the Cumont text) and a longer area beyond that, the area where the actual sacred meal took place, ending in the sacred niche and the tauroctony. The relationship between these two spaces is a Golden Ratio, such that dividing the axis of the longer second area by that of the shorter first area yields a result of 0.615 – a fairly accurate approximation of phi (the reader can satisfy himself that this is so by using a ruler with any measurement desired on the scale drawing of the Mithraeum at Heddernheim reproduced at left).

Most remarkably, as the diagram on the preceding page makes clear, the axis of the Mithraeum at Heddernheim diagrammed in the 1903 English translation of Franz Cumont's 2nd revised French edition of his *Mysteries of Mithra* is oriented 15° east of true north. This alignment matches that of the axis of the so-called Avenue of the Dead at Teotihuacan, which is also aligned east of true north. While such an alignment can be written off by skeptics as mere coincidence, it is simply another clue in the growing pile of clues from around the globe, all of which are connected by clear references to precession, precessional numbers, precessional symbology, and deliberate alignments.

At the very least, this archaeological evidence suggests that Ulansey's suspicions about the connection of Mithraic symbology to the phenomenon of precession are well-founded, although it also suggests an understanding of precession and precessional numbers both older and more precise than that of Hipparchus. In fact, the remnants of Mithraism appear to contain common codes and alignments that were passed down from whatever ancient ocean-crossing civilization influenced all the other builders of the mysterious precessional monuments around the globe.

The Hydroplate Theory and
the Mysteries of Mankind's Ancient Past

We have now spent considerable time examining the extensive evidence pointing to the likelihood that the conventional theories of anthropology and ancient history are incorrect. If this were a crime novel, the analogy would be that the authorities and the community at large have convicted the wrong suspect, and are doggedly sticking to their incorrect thesis because of some kind of prejudice or another. More evidence – perhaps some of the most shocking evidence – will be forthcoming, but first it is time to pause to lay out the alternative thesis, having already seen enough evidence both from archaeology and from the tattered remnants of human literature and myth to declare the "official story" fatally flawed.

Earlier in the book, we examined the hydroplate theory of Dr. Walt Brown, a graduate of both West Point and MIT and a former professor at the US Air Force Academy. Dr. Brown begins with one very unconventional assumption – that prior to an ancient cataclysmic global flood, there was salty water trapped in a layer underneath the earth's crust, under great pressure – and from this single assumption he proceeds to examine the evidence all over the globe which supports his theory that the violent escape of this trapped subterranean water led to the features we see on the earth today.

Dr. Brown's extensive array of evidence – all of which can be better explained by his theory than by the reigning uniformitarian geological theories, including the widely-accepted theory of plate tectonics – is almost entirely geological in nature. He does occasionally venture into anthropological evidence, such as when he examines briefly the oral traditions of the Hopi people, but in general his writing is concerned with the physical features of the earth (and the solar system) which support his explanation (and which pose tremendous problems for the currently-dominant paradigm).

It is my belief that the evidence presented so far in this book, examining in greater detail the surviving mythology from man's ancient past, as well as the surviving archaeological structures such as the pyramids of Egypt and the Americas, the stone circles, mounds, and other megalithic structures found throughout the world, and the other anthropological evidence that exists, is also explained far better by Walt Brown's theory than by the currently-accepted paradigm of geology. In other words, the anthropological evidence provides a powerful support of his theory, and is explained better by his theory than by some of the other alternative theories which those who have also noticed the anomalous evidence tend to put forward (alternative theories such as the idea that the planet Venus brushed close to our earth in the distant past, causing massive crustal displacement that left the geological features we see today).

Dr. Brown's theory flies in the face of uniformitarian geologic theories that have dominated academia since the 1800s. Uniformitarian theory is so named because it posits that all or almost all of the features we find in the geology on earth today can be explained by processes which are for the most part still going on – in other words, processes that are "uniform," as opposed to catastrophoc. It looks for the same forces going on today but acting over long periods of time, rather than unusual forces which shaped the earth with processes dramatically unlike those we still see in operation today.

The introduction of uniformitarian geological theories in the early 1800s was almost as revolutionary as the introduction of Darwinian theories of evolution a few decades later, and in fact the two theories are closely related and their principal proponents knew each other well and discussed their mutual work together in person. Uniformitarian theories had been proposed earlier than the 1800s, most notably by pioneering Scottish geologist James Hutton (1726 – 1797), who published two texts entitled *Theory of the Earth* and *Concerning the System of the Earth* in 1785. In these works, Hutton outlined his arguments that processes like those going on today could have shaped the features we find today, given enough time (and he proposed vast eons of time stretching so far back that we can today find "no vestige of a beginning, -- no prospect of an end").

This suggestion was a radical departure from the prevailing theories of the day, which were (at least in Europe, the United States, and other Western nations) based primarily upon the Biblical account in Genesis, relying upon the forces of the Biblical Flood (this is the source of the term "antediluvian," or pre-Flood, for anything considered extremely ancient) and positing a relatively young earth created some thousands or perhaps tens of thousands of years ago, rather than the endless millions posited by Hutton.

While Hutton's work laid an important foundation for uniformitarianism, it was really the work of Charles Lyell (1797 – 1875) which vaulted uniformitarianism into the popular consciousness, primarily with the publication of his *Principles of Geology: an attempt to explain the former changes of earth's surface by reference to causes now in operation* (1830), a book which would be so popular that it supported twelve more editions over the course of Lyell's life (the twelfth being published posthumously in 1875). His other major works included *Elements of Geology* (1838), *Travels in North America* (1845), and *Geological Evidences for the Antiquity of Man* (1863).

The very first quotation in the first edition of *Principles of Geology* was a quotation which reads:

> Amid all the revolutions of the globe the economy of Nature has been uniform, and her laws are the only things that have resisted the general movement. The rivers and the rocks, the seas and the continents have been changed in all their parts, but the laws which direct those changes, and the rules to which they are subject, have remained invariably the same. Playfair, *Illustrations of the Huttonian Theory.*

This quotation can be seen as a uniformitarian manifesto – the declaration that all the changes which have worked over the earth are the result of natural forces that are unvarying: uniform. It is simultaneously a rejection of the idea of extraordinary or even supernatural forces resorted to by churchmen of the time to explain the features of the earth around them. A more succinct encapsulation of this fundamental uniformitarian assumption is Lyell's own phrase, "The present is the key to the past."

Uniformitarian theories inevitably require extremely long periods of time. If features as enormous as the Grand Canyon in Arizona, for instance, were created not by catastrophic forces but instead by the same forces of erosion that the Colorado River is exerting today, then tens or even hundreds of millions of years must be added so that those much milder uniformitarian forces can have any hope of doing the job.

Conversely, catastrophic theories undermine the need for vast ages which have dominated the prevailing theories since Hutton and Lyell. This is a very important point. If enormous geological features such as the Grand Canyon could have been formed by catastrophic forces in relatively short periods of time (in other words, not requiring millions of years but perhaps only months or even weeks), then the need for endless rolling millions of years marching off into unknowable antiquity is seriously compromised. If the Grand Canyon and other natural features could have been formed in weeks or months, then it is possible that they were formed last month – or rather, since they are populated with trees and bearing some signs of age, at least only a few thousands of years ago instead of millions.

It should be obvious, then, that uniformitarian geology, by its reliance upon and assertion of endless millions of years of time, is a strong and natural ally of Darwinian evolution, which also relies upon vast amounts of time to produce mutations as complex as spiders spinning webs that are stronger than an equal volume of steel and yet far more flexible, or bats which can negotiate a dark room strung with a complex tangle of piano wire at high speeds, and unerringly catch a flying moth in the process.

And in fact Darwin and Lyell were friends, sharing long conversations about their respective theories. Lyell was reluctant to accept all of the implications of Darwin's theory, particularly Darwin's assertion that man is descended from apes and the specifics of natural selection, but the two were close friends and there is some evidence from Lyell's writings over the decades that he became more and more amenable to various aspects of evolutionary theory, even if he never fully accepted Darwinism in totality.

The obvious synergy between the geological theories of uniformitarianism and the biological theories of Darwinism mean that any geological theory which challenges uniformitarianism (as Dr. Brown's hydroplate theory clearly does) will be violently attacked not only by those who have a stake in preserving geological uniformitarianism but also by those who are devoted to Darwinian evolution. Any threat to theories which require hundreds of millions of years of geological shaping will necessarily pose a threat to the hundreds of millions of years required by Darwin's theory as well. Like the authorities in a crime novel who are motivated by their own prejudices to accept an explanation that is obviously flawed, the adherents of the conventional worldview are predisposed to explain any evidence from the perspective of their own assumptions, and to reject explanations coming from assumptions which threaten their own core beliefs. This bias can operate almost entirely below the level of conscious thought, so that those who misinterpret the evidence think they are being completely impartial, much like a referee at a sporting event who believes he is making an unbiased call but is really being unintentionally harder on one team for some reason.

When first encountered, Brown's hydroplate theory may be difficult to accept as well because of the startling nature of his first assumption, that water had been trapped beneath the crust under high pressure prior to the global flood – so much water, in fact, that its volume makes up roughly half of what is now in the oceans. Remember, however, that numerous flood legends which remain to us as artifacts from humanity's past seem to include water coming up from below to flood the earth. Most notable of these, of course, is the Genesis passage which relates that "In the six hundredth year of Noah's life, in the second month, the seventeenth day of the month, the same day were all the fountains of the great deep broken up, and the windows of heaven were opened" (Genesis 7:11).

The order here is interesting, in that it was the "fountains of the great deep" (whatever that means) which were broken up, followed by the opening of "the windows of heaven." According to Brown's theory, it was the rupture of the crust which enabled the explosive escape of the water under the earth, some of it jetting as high as twenty miles into the atmosphere, which was then followed by torrential downpours as the water came back down.

Another passage from the Hebrew Scripture worth noting is that of Job chapter 38, in which the Almighty finally answers Job with a string of questions that leave Job in awe-stricken silence:

> "Then the LORD answered Job out of the whirlwind and said, Who *is* this that darkeneth counsel by words without knowledge? Gird up now thy loins like a man; for I will demand of thee, and answer thou me. Where wast thou when I laid the foundations of the earth? declare, if thou hast understanding. Who hath laid the measure thereof, if thou knowest? or who hath stretched the line upon it? Whereupon are the foundations thereof fastened? or who laid the cornerstone thereof; when the morning stars sang together, and all the sons of God shouted for joy? Or *who* shut up the sea with doors, when it brake forth, *as if* it had issued out of the womb? When I made the cloud the garment thereof, and thick darkness a swaddlingband for it, and brake up for it my decreed *place*, and set bars and doors, and said, Hitherto shalt thou come, but no further; and here shall thy proud waves be stayed? Job 38:1-11.

Here we find the rather startling imagery of the sea being described as breaking forth as if "it had issued out of the womb" (when it burst forth out of its boundaries, before later being confined to its "decreed place" which the Almighty "brake up for it"). We have already seen that the legends of the Cuna tribes of what is today Panama in Central America describe the flood as "being due to breaking of the great water jars of the underworld by a god" (Keeler 60).

Brown's hydroplate theory involves the eventual runoff of the floodwaters into the great basin of the Pacific and Atlantic, and according to his theory the Pacific basin was in fact "broken up" by the chain of events set into action by the original rupture: the water jetting out eroded tons of earth from the sides of the original crust, widening the crack and reducing the weight above until the basement rock that had been below sprang upward (and producing the massive volumes of sediments which would be sorted by the water's action into the sedimentary layers found all over the earth today).

This upward springing of the basement crust caused the initiation of the sliding of the "hydroplates" (still lubricated below by the water that had not escaped) away from what is today the Mid-Atlantic Ridge. At the same time, this action also caused an oppo-

Image from the National Oceanic and Atmospheric Administration (NOAA), clearly showing the Mid-Atlantic Ridge and the continental shelves which (according to the hydroplate theory) were formed by the violent escape of massive flows of water.

site reaction on the other side of the globe as the upward movement of mass on the Atlantic side of the globe caused a giant downward suction of mass on the Pacific side, breaking up the crust and plate that were there and creating the violent cusps around the depression that resemble more than anything the crater that is caused if a man presses his thumb into the side of a ping-pong ball. The hydroplates slid toward this new abyss, eventually grinding to a stop due to friction, buckling in the process like the front of a big truck when it hits a wall. Along their front edges in particular, the incredible heat and friction produced magma as the plates themselves melted underneath, which explains the volcanic formations found around the edge of the Pacific on all sides. The violent breaking up of the crust that now forms the ocean floor of the Pacific also released magma and heat, which is why the "Ring of Fire" can still be traced around the edge of the Pacific. Today, although the sliding action has largely ceased, there is ongoing shifting and settling of the mighty plates towards the direction of the Pacific abyss, resulting in earthquakes – which remain most common in the Pacific and the continents around its boundaries.

Difficult as such a paradigm-shifting theory may be to accept among those of us who have been conditioned our entire lives to believe in uniformitarian geology (by educators at all levels), Brown's hydroplate theory appears in many ways to have greatly superior explanations for the evidence we find around the world, including not only the arc-and-cusp shapes of the trenches in the Pacific, but also the salty water found by miners when they dig very deep shafts miles down into the crust, or the remains of masses of temperate trees and vegetation in the very high latitudes on the globe which today could not possibly support growth of that size and volume (and are today covered with permafrost and support only the barest lichens and mosses). As we have already seen, Brown's theory posits that the rapid buckling that led to the most massive mountain ranges on earth (the Himalayas and surrounding chains) caused a change in the actual center of gravity of the globe, such that it rolled between 35° and 45°, bringing the previous north pole of the earth down to the region of Manchuria, and bringing regions that once were in more temperate latitudes up to the new north pole (and down to the new south pole on the other side of the sphere as well).

Brown's theory, in fact, explains literally hundreds of geological phenomena on earth in ways that he shows to be much more plausible and in line with the principles of physics than the prevailing explanations. One prominent example is his theory's explanation of the formation of the Grand Canyon (that outstanding laboratory for geology and geological theory). The uniformitarian theory, of course, holds that the enormous Grand Canyon was carved out by erosion through the action of the Colorado River over millions of years. This is what park rangers confidently tell visitors, and what numerous official-sounding movies and geographic specials state as settled fact, without the slightest hint that anyone could ever doubt such an explanation.

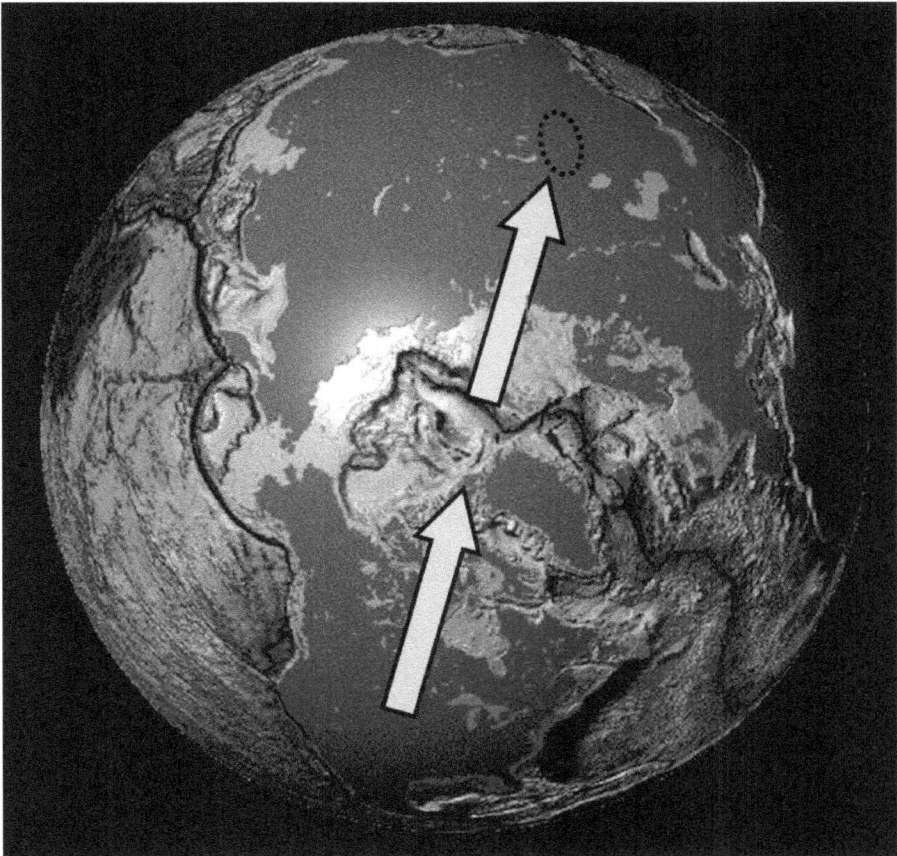

The roll of the earth caused by the thickening of the Asian plate at the Himalaya region. This image looks down on the current north pole, which rolled up to its present location from the bottom of the lower arrow. The former north pole rolled to the area indicated by the dotted oval, which moved along the path indicated by the upper arrow.

In fact, however, this explanation is ludicrous, and runs into numerous king-sized problems when it is forced to account for the actual evidence on the ground. For starters, Dr. Brown notes that all conventional theories have to explain how the Colorado River, which flows generally southward on the eastern side of the high Kaibab Plateau, suddenly took a hard right turn to the west and flowed up and over that plateau, eroding a mighty canyon right through the middle of it. The diagram below makes this problem clear.

Dr. Brown points out:

> All explanations for the Grand Canyon's origin try to answer this question. Some say that the river was once a mile or

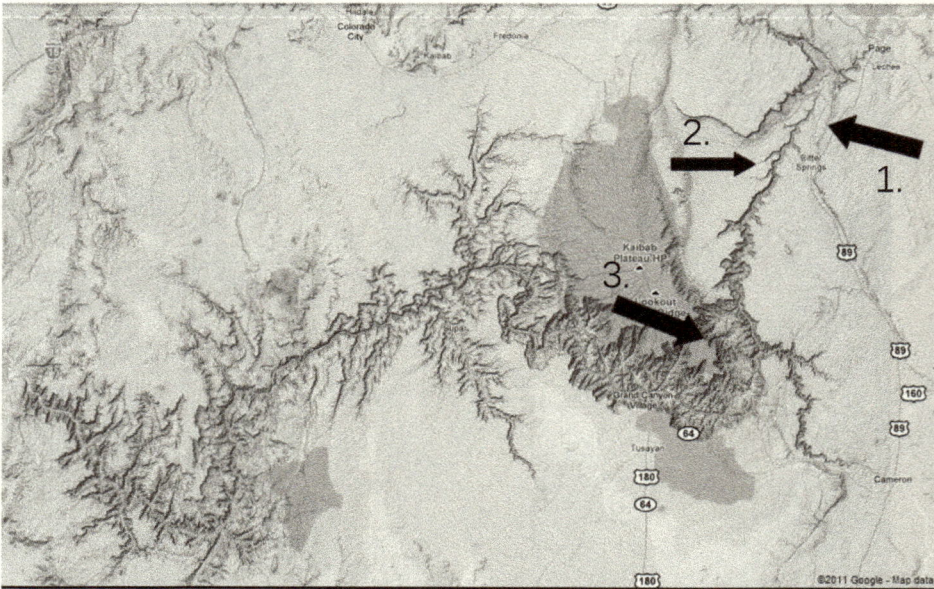

Google Map with terrain features showing the distinctive and extreme terrain of the Grand Canyon. This view is particularly useful because it clearly illustrates the obvious "funnel" blasted out of the high cliffs above Bitter Springs (arrow 1). The distinctive "barbed canyons" emanating from the direction of water travel along the Colorado River are clearly seen near the funnel feature. One such barbed canyon is marked with arrow 2. Tributary streams do not typically enter a larger river from this direction -- undermining the conventional theory that a normal river carved the Grand Canyon over millions of years. Finally, the conventional theories have tremendous difficulty explaining why and how the Colorado River turned and flowed right through the massif of the Kaibab Plateau (at arrow 3).

more higher, and the land it flowed over eroded away. As it did, the river settled down on top of the Kaibab Plateau and cut through it – a process called *superposition*. Others say the river cut through the Kaibab Plateau along a fault (or crack). However, faults [in the Kaibab Plateau] are generally perpendicular to the Colorado River, not parallel. Some believe that the land under the river rose, forming the Kaibab Plateau. As it did, the river cut down through the rising plateau. Two theories say that a stream flowing down a western slope of the Colorado Plateau continually eroded *eastward* 130 miles and eventually cut through the Kaibab Plateau – a process called *headward erosion*. (Notice how dependent these explanations are on millions of years of time, and how many untestable explanations can be proposed if millions of years are imagined). *In the Beginning,* 8th ed. 187ff.

Other features Brown discovered that provide powerful support for his explanation include the massive "funnel" shaped feature

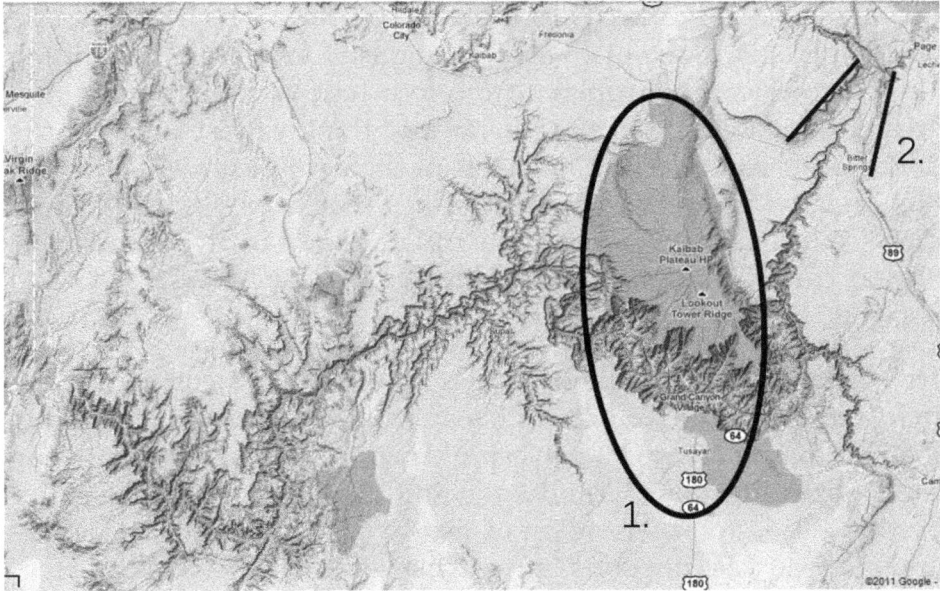

To ensure the reader sees the funnel and the Kaibab Plateau being described in the text and in the previous diagram, the same terrain map of the Grand Canyon is reproduced here. This time, the Kaibab Plateau is marked by the general shape of the ellipse at 1. The funnel feature above Bitter Springs is marked by two lines at 2.

that can be clearly seen in the map at right, as well as the "barbed canyons" which flow in the opposite direction to the direction water typically flows into a river.

The features are shown again in the image below, with the Kaibab Plateau outlined in the large ellipse, and the funnel shape feature outlined with diverging lines.

Dr. Brown piles on other additional clues from the Grand Canyon that all point to the hydroplate theory as the best explanation, evidence which give conventional explanations considerable difficulty. Among these are the features of Marble Canyon, Nankoweap Canyon, and many aspects of the famous layering visible in the Grand Canyon itself (among them the question of why erosion would cut down so deeply into harder and harder layers of rock, instead of eroding primarily horizontally in softer upper sedimentary layers before carving down deeper).

Perhaps most striking is the evidence from the regions to the east and northeast of the Grand Canyon, where Brown proposes the enormous lakes lay before breaching their western wall and emptying their contents. Brown suggests that after the sliding North American hydroplate came to a halt (a violent event that created the buckling of the Sierras and the Rockies), much of the floodwaters ran off into the basins of the Atlantic and Pacific, but huge reservoirs were trapped in areas that later became California's Central Valley, the salt flats of Utah, and in two large lakes Brown calls Grand Lake and Hopi Lake east of today's Grand Canyon.

The sinking action of the heavy Rockies created a corresponding rising action of the Colorado Plateau (just as the sinking action of the Himalayas created a corresponding rising action that formed the Tibetan Plateau). Grand Lake and Hopi Lake were raised up by this action to an elevation of nearly 6,000 feet above sea level, and covered vast portions of the American Southwest, with Hopi Lake stretching across much of northeast New Mexico and the larger Grand Lake stretching east to the Four Corners region and northeast across the entire southeast corner of Utah and into Colorado.

Brown explains how this theory solves not only the mystery of the Grand Canyon's origins (and its many features that conventional theories cannot satisfactorily explain) but also the many amazing geological wonders of the American southwest, including the gathering of huge petrified logs (jumbled in shattered heaps, which is difficult to explain with other theories but which can be readily explained if those logs were petrified at the bottom of a warm mineral-laced inland sea which then drained rapidly, sucking them violently into a corner of the erstwhile lake and moving with enough force to shatter even those huge stone logs), the majestic mesas and buttes and spires of Arizona's Monument Valley (featured in numerous Western movies and also explained by the rapid draining of Grand Lake, within whose ancient boundaries these remnants are primarily found), the incredible features of Bryce Canyon (located along the ancient edge of Grand Lake and also formed by the rapid emptying once it breached), and many others.

Dr. Brown's examination of the evidence at the Grand Canyon (which we have only touched on here in cursory fashion) is just one of many examinations that he performs on terrain features from around the world, all of which can be better explained by the impact of a catastrophic flood and its after-effects than by conventional theories.

Others include:

- The fossil evidence of mighty forests and swamps on Canada's Ellesmere Island (located north of 79° north latitude), where very little vegetation can survive today. As we have already seen, Brown's theory argues that the rapid creation of the Himalaya region during the compression event that ended the sliding of the hydroplates caused a roll in the earth such that the previous north pole rolled about 35° to 45° in that direction, and that modern-day Mongolia was once at the north pole. Brown finds other supporting evidence that this roll of the earth took place, primarily the existence of the undersea ridge known as 90 East Ridge (88, 117).

- The frozen mammoths found preserved in recent centuries in northern Siberia and Alaska. Brown explains that, contrary to depictions in many children's books of prehistoric animals, mammoths could not have been arctic creatures. The food and water needs of modern elephants (330 pounds of forage and 30 to 60 gallons of water every *day*) would be nearly impossible to obtain in the frozen tundra. The stomachs of intact mammoths frozen in the far north contain ripe fruits, herbs, shrubs, tree leaves, bean pods, and flowers such as buttercups – most of which are very difficult to explain. How such massive animals could have obtained sufficient food and water especially during the dark Arctic winters is a major problem for conventional theories. Further, elephants cannot sustain extreme cold temperatures, primarily due to the massive heat loss through their long trunks. Brown provides several other arguments pointing to the conclusion that mammoths were not originally Arctic animals but rather lived in temperate latitudes prior to the flood event. Even more difficult to explain is the fact that these huge animals were frozen so completely and so rapidly that not only was their flesh, skin, and hair preserved, but also the contents of their stomachs. Dr. Brown's theory explains that some of the jetting water and sediments from the world-encircling rupture would have achieved high altitudes and frozen, falling as tons of muddy ice crystals and hail, suffocating and freezing many animals very rapidly. The roll of the earth discussed previously would have carried the mammoths from their previous temperate latitudes to the far north. In Dr. Brown's words, "As the flood waters drained off the continents, the icy graves in warmer climates melted, and the flesh of those animals decayed. However, many animals, buried in what are now permafrost regions, were preserved" (169).

- The formation of the sediments and fossils themselves. The hydroplate theory argues that the sedimentary layers were the product of the masses of eroded material from the original rupture, which were then sorted into layers by the phenomenon of liquefaction, in which sediment particles are surrounded by water and begin to behave more like a liquid than a solid. This process sorted both the sediments and the rapidly buried animals into the layers that are today wrongly interpreted by conventional Darwinian and uniformitarian assumptionsas representing various epochs of millions of years each.   The conventional assumptions have been repeated so many times that most people accept them without much critical thought, but they contain many obvious problems.   As Dr. Brown points out, the creation of fossils requires very unusual circumstances, since dead animals and plants are typically eaten or else decompose fairly quickly, leaving no fossils behind.   Fossilization requires rapid burial in something that can prevent bacterial decay – for instance, rapid burial in thick moist mud.   The fossil record contains abundant evidence of rapid burial – for instance, the numerous fossils of fish caught in the act of swallowing another fish. The numerous examples of such fish-eating-fish fossils is consistent with almost instantaneous death and burial, rather than normal causes of death.   Further, Brown notes that fish fossils are often found "flattened between extremely thin sedimentary layers.   This requires squeezing the fish to the thinness of a sheet of paper without damaging the thin sedimentary layers immediately above and below" (143). He notes that even dumping tons of sediment through water onto fish would not produce such

National Park Service image of a fossil of a fish swallowing a fish (many other images of fish preserved in the act of swallowing another fish exist -- this is by no means a unique or isolated example). For a fossil to be preserved at all, the animal or plant must be rapidly buried in a sealant (such as heavy mud) to prevent being eaten by scavengers or bacteria.   Note that fish fossils are often paper-thin, which is consistent with the explaration of Walt Brown's hydroplate theory but very difficult to explain using most conventional theories.
image: Fossil Butte National Monument, Wyoming

an effect, but liquefaction explains it quite well (143). It also explains the fossilized outlines of soft, rapidly-decomposing animals such as jellyfish.

- The nature of the sedimentary layers and evidence that they were soft when they buckled. Dr. Brown observes that: "Sedimentary layers usually have boundaries that are sharply defined, parallel, and nearly horizontal. These layers are often stacked vertically for thousands of feet. If layers had been laid down thousands of years apart, erosion would have destroyed this parallelism" (142). The very nature of the sedimentary layers – particularly their sharp edges and well-defined parallelism – argues for their rapid creation, rather than their production via millennia as in conventional theories. Further, there are many places on earth where these strata are violently buckled but not shattered. If they had been laid down over successive millions of years, most of the strata would have been hard

Folded sedimentary rocks at St. Anne's Head, Wales. Folds like these indicate that the sediments were still pliable when the force was applied. Today they are brittle. This accords with the hydroplate theory. Examples similar to this can be found around the world.
image: Wikimedia commons. Photograph: Rodney Harris.

and brittle. However, the folding patterns we can see today suggest that the strata were soft and pliant when bent into their current shapes, and later hardened (see images below for example; there are many more around the world).

There are literally hundreds of other geological examples in Dr. Brown's work, from volcanic activity to comets, and I would encourage the reader to explore them in the original, but at this point it is worthwhile to consider whether the hydroplate theory can explain the extensive evidence we have considered prior to this chapter from man's ancient past.

At this point, having read the assertion that the strata were laid down in a fairly recent cataclysmic event (10,000 years ago or less), some readers may be wondering about radiocarbon dating, which purports to find great ages on certain ancient artifacts. We have examined radiocarbon dating already in some detail, in conjunction with the dating of the skull of Ruamahanga Woman. From that discussion, it is clear that radiocarbon dating can only measure carbon-14 in things that were once alive, since the dating method compares the amount of unstable carbon-14 in the sample to the amount assumed to be present in the atmosphere when the sample was part of a living organism absorbing carbon from the atmosphere (or by eating things which had absorbed carbon from the atmosphere).

The key word in the foregoing sentence is "assumed" – radiocarbon dating depends upon an assumption of how much carbon-14 was in the atmosphere at remote dates in the past. In fact, it assumes that the level of carbon-14 is fairly constant, and that there was as much in the atmosphere tens of millions of years ago as there is today (with some fluctuation due to the advent of the industrial age and the detonation of powerful nuclear bombs during the 1960s).

Walt Brown's hydroplate theory, however, challenges this assumption, noting that a massive eruption of underground water would release vast amounts of carbon dioxide but that this new carbon would contain little or no radiocarbon (carbon-14) due to the fact that it had been underground and not subjected to cosmic and solar radiation. Afterwards, the ratio of carbon-14 to carbon-12 would rise rapidly (assuming that carbon-14 forms more rapidly than it decays, based on its long half-life). This rapid rise would eventually taper off to a relatively constant ratio, where it is today. Dr. Brown provides arguments that this equilibrium was achieved about 3,500 years ago, which means that radiocarbon dating back that far will be relatively accurate.

However, prior to reaching this level of equilibrium, the actual levels of radiocarbon in the atmosphere were much lower than radiocarbon dating assumes. This means that specimens that lived prior to 3,500 years ago would have absorbed lower levels of carbon-14 during their lives, and that radiocarbon dating will yield an erroneously higher age when analysts detect the remaining carbon-14. Using their assumption of constant carbon-14, they will assume that many more half-lives of decay have taken place to reduce the levels of carbon-14, when in fact the organism did not absorb the carbon-14 during its lifetime in the first place. Because the increase in carbon-14 was rapid prior to reaching equilibrium, this error gets increasingly larger the further back one goes. Also, because scientists are using the half-life of carbon-14 in their calculation, the erroneously old dates that they calculate will grow exponentially the older the specimen actually is.

Thus, radiocarbon dating – while valuable for dating organic material from 3,500 years ago or less – is not a valid counterargument to the hydroplate theory, because if the hydroplate theory is correct, then the levels of radiocarbon in ancient atmospheres would have been vastly lower.

One of the important aspects of Dr. Brown's theory is his description of the chain of events that led to the filling of the oceans. After the rupture and flood phases of his theory, the hydroplates began sliding away from one another after the basement rock sprang upwards in response to the removal of massive amounts of sediments by the jetting escaping waters. When they buckled and thickened due to friction and the laws of physics, the floodwaters poured off into the deep ocean basins (carving massive canyons in the sides of the continental slopes that remain to this day but which are very difficult for conventional theories to explain).

The oceans at that time were much lower relative to the continents,

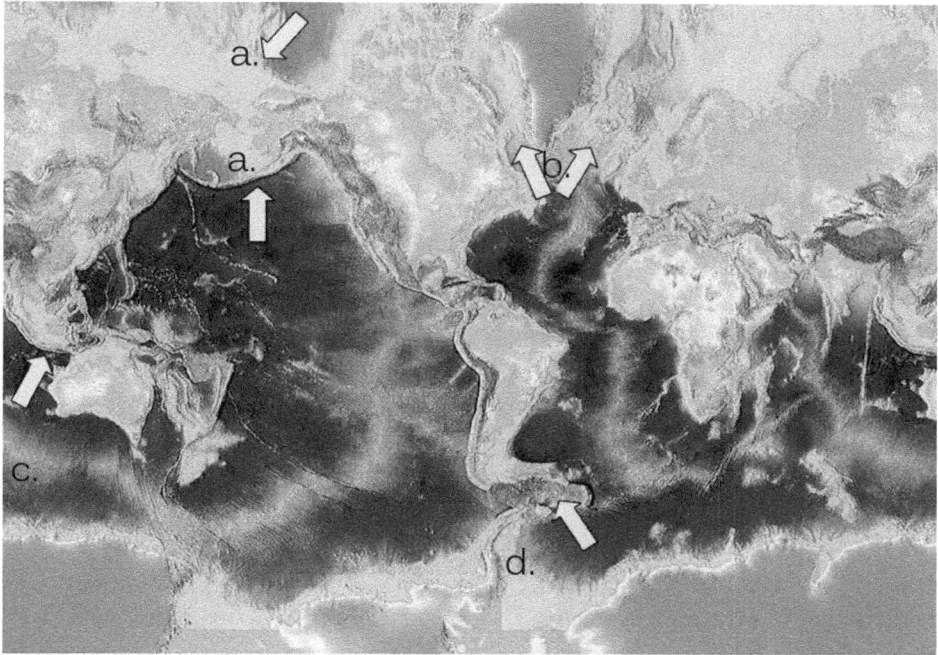

NOAA map of the earth showing ocean floor depths. The hydro-plate theory argues that after the initial rupture and flood event, the continents accelerated away from the rupture, then compressed and thickened when they ground to a halt due to friction or collision. Floodwaters drained down the steep continental slopes into the ocean basins. For several hundred years, ocean levels were much lower than they are today, and continents were much higher rela-tive to the oceans, exposing land bridges between the continents, at the Bering Strait between Siberia and Alaska (arrows a.), between Europe Greenland and North America (b.), between Australia and Southeast Asia (c.), and even between South America and Antarctica (d.). In some cases, a small channel of water might have been pres-ent as a minor obstacle, but in general, land migration was possible.

which had only recently thickened and had yet to sink under their own weight downwards into the basement rock below. Because of this fact, several important land bridges connected the continents, as shown in the diagram above. Men and animals were able to migrate between continents (and to the Galapagos Islands, which at that time were a peninsula extending from South America, which would later be swallowed up as the seas rose, leaving only the isolated tips protruding as the islands we see today).

However, due to the laws of physics, the thickened continents slowly sank downwards, causing the ocean floors to rise and the ocean levels to become much higher relative to the land. Also, large trapped inland seas which we have discussed earlier in conjunction with our examination of the Grand Canyon breached over the centuries (due to rainfall and icemelt) and dumped their contents back into the oceans.

Dr. Brown explains the physics of this sinking of the continents and rising of the seas: "Because the thickened hydroplates applied greater pressure to the floor than the water, the hydroplates slowly sank into the basalt floor over the centuries, causing the deep ocean floor to rise. (Imagine covering half of a waterbed with a sheet and the other half with a thick metal plate. The metal plate will sink, causing the sheet to rise)" (106).

Thus, the hydroplate theory provides a clear explanation for lowered ocean levels immediately after the flood followed by rising ocean levels over the next few centuries which eventually covered the large land bridges and also submerged islands which once were above water, as well as much of the land around existing islands and island chains including the Maldives, the Galapagos (which was likely once a large peninsula extending from the coast of Chile), Malta, Japan, Cuba and elsewhere.

This theory explains – in ways that conventional theories do not – the evidence of human architecture standing on the ocean floor at depths of hundreds of feet. Much of these underwater cities have only recently been discovered, and more are being discovered each year. The guardians of the conventional framework often feel very threatened by these discoveries and do their best to ignore or discredit them.

The prime example of undersea megalithic ruins is the site located off the shores of the island of Yonaguni near Okinawa, Japan. The site was not discovered until 1995, when a diver accidentally strayed off course and saw huge blocks lying in apparently regular patterns at depths of forty feet. In his book *Underworld*, Graham Hancock chronicles his own dives at the site, as well as at other less well-known sites around the world, particularly off the coast of India at depths of over a hundred feet of ocean water.

The hydroplate theory's assertion that the ocean levels were once far lower and then rose with the sinking action of the continents also explains the migrations of peoples across the vast abyss of the Pacific to the Americas. It of course provides a mechanism whereby the land bridge across the Bering Strait would have been above sea level, which is more plausible than the conventional argument that the reduction of sea levels is due to ocean water being trapped as ice in previous Ice Ages, but it also provides an explanation as to how the ocean itself could have been crossed even by peoples who did not have the nautical ability to circumnavigate the globe. If the ocean levels were much lower, there would have been more islands above the surface to provide "stepping stones" for island hopping across the Pacific.

As Dr. Brown's book points out in a footnote on page 255 of the 7th edition, the oral history of the Hopi people supports just such

Undersea ruins at Yonaguni. This formation is known as "the Turtle." Defenders of the conventional orthodcxy are at pains to argue that these formations are the product of natural wave action.
Image: Wikimedia commons Photograph: Masahiro Kaji.

an explanation for the original migration of their ancestors across the ocean from the east to the west. In the *Book of the Hopi* (1962), Frank Waters (1902 – 1995) recounts the worldview and history of the Hopis of northern Arizona, as it was told to him by thirty elders of the tribe.

The account first tells of the creation of mankind – interestingly, the creation account explains that Taiowa, the creator deity who is also associated with the sun, created a nephew named Sótuknang (note the telling linguistic similarity between the name Taiowa and the covenant name of the Creator revealed in the Hebrew Scriptures). This nephew then began to order the universe and created someone to help him, Kókyangwuti or Spider Woman. When these two had created a world ready for human life, Spider Woman proceeded to make mankind. The account is very interesting, for it shows awareness of the various races of mankind, which seems anomalous for an ancient people living in the desert of Arizona. The account as recorded by Frank Waters is as follows:

> So Spider Woman gathered earth, this time of four colors, yellow, red, white, and black: mixed with *túchvala*, the liquid of her mouth; molded them; and covered them with her white-substance cape which was the creative wisdom itself. As before, she sang over them the Creation Song, and when she uncovered them these forms were human beings in the image of Sótuknang. Then she created four other beings after her own form. They were *wúti*, female partners, for the first four male beings. 5.

Clearly, either the legends of the Hopi were altered after Europeans arrived on the scene after the time of Columbus and the Hopi were exposed to other races, or else the Hopi have preserved with great accuracy the knowledge of the existence of people of other races from before the arrival of Columbus (possibly from pre-Columbian visits from other continents). The only other possibility is that the Hopi legends just randomly tell of people being made in four races of four colors, even though they had never themselves met anyone of other races, and that they just happened to coincidentally select the colors yellow, red, white and black.

That different races are being described by this creation account is made clear by later passages, such as the account in which Spider Woman and her helpers (two twins who are stationed at the north pole and south pole to keep the world properly rotating on its axis!) summon Sótuknang to help give the first people the power of speech. The account relates: "So Sótuknang gave them speech, a different language to each color, with respect for each other's difference" (7). Later in the account, a disruptive being comes in the form of a bird called Mochni (who is like a mocking bird) whose deceptive talking began to convince the first people of differences between them – "the difference between people and animals, and the differences between the people themselves by reason of the colors of their skins" (12). Still later, a deceptive and handsome serpent shows up who further leads them away from the wisdom of their creation state and stirs up more suspicion between different people, leading now to actual violence and warfare (12).

Even more amazing, the Hopi account tells of the destruction and recreation of three successive worlds, just as the legends of the ancient Maya and ancient Hindu legends do. Frank Waters records that the Hopi elders recounted the destruction of the first world by fire, the second by ice, and the third by water. The Hopi believed that they were now living in the fourth world, just as other cultures also believed. The account of the destruction of the second world by ice is particularly noteworthy: just as before, the people who are to escape destruction are led underground to the kiva of the Ant People, and then:

> When they were safely underground, Sótuknang commanded the twins, Pöqánghoya and Palongawhoya, to leave their posts at the north and south ends of the world's axis, where they were stationed to keep the earth properly rotating. The twins had hardly abandoned their stations when the world, with no one to control it, teetered off balance, spun around crazily, then rolled over twice. Mountains plunged into seas with a great splash, seas and lakes sloshed over the land; and as the world spun through cold and lifeless space it froze into solid ice. 16.

The connection of destruction with the unhinging of the world-axis is stunning. The understanding of the world as a globe is equally startling. The connection with legends we have seen from around the world is more astonishing still.

It is after the destruction of the Third World by water that the account of the Hopi people coming to their present home takes place. First, Sótuknang instructs Spider Woman to seal up the people who are to be saved inside a vessel made of hollow reeds. Then: "he loosed the waters upon the earth. Waves higher than mountains rolled in upon the land. Continents broke asunder and sank beneath the seas. And still the rains fell, the waves rolled in" (18). Then, they landed upon a little piece of land that had once been the top of one of the highest mountains. Led by Spider Woman, the people make rafts of hollow plants and sail to one island after another, leaving their rafts at landfall and traveling by foot eastward to the other side, only to be told they must make rafts again and continue "east and a little north" (18-19). Over and over this goes on, through rich islands filled with seed-bearing and nut-bearing trees, but each time Spider Woman tells them that this land is not for them. At last, they reach "a great land, a mighty land" with towering walls so steep they cannot land, "stretching from north to south as far as they could see" (20). They go north, they go south, but they cannot find a break in the mighty walls (Dr. Brown's theory explains that these were the slopes of the continental shelf, which are now under the ocean). At last, they "stopped paddling, opened the doors on the tops of their heads [the elders earlier explained that man has "several vibratory centers" along the axis of his body, just as the "living body of the earth" does; these correspond remarkably to the *chakras* of Hinduism, and the highest and most important of these is at the crown of the head, see Waters 9-10] and let themselves be guided" (20). The elders recount that when they did this, they were swept up in a gentle current that deposited them at last in the Fourth World, their new home.

Sótuknang directs them to look back, and they are able to see "sticking out of the water the islands upon which they had rested" on their long journey. The elders recount:

"They are the footprints of your journey," continued Sótuknang, "the tops of the high mountains of the Third World, which I destroyed. Now watch." As the people watched them, the closest one sank under the water, then the next, until all were gone, and they could see only water. 20.

This account is remarkable, all the more so because it accords very closely with the mechanics proposed by the hydroplate theory. As Brown explains in his footnote, he was not originally aware of the Hopi legends – a reader brought the close correlation to his attention after Brown had originally published his theory (255). The existence of a myth or oral tradition (a myth of which Colonel Brown was unaware when he first published his theory) which appears to confirm the assertions of his theory must be counted as yet another "data point" in the hydroplate theory's favor.

There are more fascinating connections to be found in Waters' priceless account of the traditions of the Hopi as related to him by the elders themselves some fifty years ago, among them the prominent place held by the Parrot clan, which "began their migration in the warm country far to the south," and the history of the Condor clan, which according to legend had moved along the high mountains of South America and there "built up a great city of stone" (54-56). Also noteworthy is the belief among the modern Hopi that the serpent mounds found in Ohio and the eastern US might have been built by their ancestors, as the symbology of a serpent swallowing an egg is very familiar to them (40-41). The clues left to us from the ancient traditions of the Hopi are all consistent with the other clues we have seen in other parts of the world, namely that there were trans-oceanic migrations long before traditional recorded history, that ancient people had knowledge of the shape of the earth and facts about its rotation, that an ancient cataclysm or cataclysms were known to man and recorded in his oral histories, complete with the connection of such cataclysms to the unhinging of the earth axis, and that there was some connection between the civilizations that built the ancient monuments that still stand at important archaeological locations around the globe today. All of these clues support the explanation of events in the hydroplate theory, and cast doubt upon uniformitarian theories that rely on hundreds of millions of years and slow processes acting long before the arrival of mankind (including the tectonic theory).

Another important feature of the hydroplate theory in regard to human anthropology is its explanation of the sedimentary layers forming via a mechanism not involving tens or hundreds of millions of years between each layer. Many alternative archaeologists or authors who have questioned the conventional paradigm nevertheless accept the uniformitarian geological framework and its explanation of the ages of the strata. Because of this, when human artifacts are found in strata that the conventional timeline dates at tens of millions or hundreds of millions of years ago, these alternative authors leap to the conclusion that modern man must have been around tens or hundreds of millions of years ago.

For example, the groundbreaking 1995 work of Michael A. Cremo and Richard L. Thompson, *Forbidden Archaeology*, provides massive evidence and documentation of human fossils, footprints and artifacts in geological layers commonly dated as hundreds of millions of years in age. While applauding their willingness to boldly challenge the conventional framework (and endure the withering counterfire of the rabid defenders of Darwinian evolution), if the accepted paradigm for the age of the strata is seen to be built upon a fabric of incorrect assumptions, then the voluminous evidence in Cremo and Thompson's 914-page work can be seen as further anthropological support for the hydroplate theory. Instead of arguing for an emergence of mankind in present form hundreds of millions of years ago, the "anomalous" fossil evidence instead argues for a strata-forming catastrophe that took place while man was on the scene – perhaps only thousands of years in the past.

This explanation accords with the existing Native American legends which describe the carving of the Grand Canyon during the days of their ancestors, or the ancient Hindu scriptures describing the formation of the Jhelum River gorge from the discharge of an ancient lake in the Kashmir Basin, as Dr. Brown describes in his book (110). As Brown explains, modern science has recently concluded that the Kashmir Basin was once filled with water. The existence of an ancient record stating such long before modern science reached the same conclusion could be mere coincidence, or it could be further evidence that man had begun to repopulate the earth after the global flood by the time these mighty inland seas finally breached. These same humans could also be responsible for the evidence Cremo and Thompson cite in their work.

It would not be surprising to find ancient human archaeological evidence buried under the ice of Antarctica in the future, if the hydroplate theory is correct and the earth experienced a great roll after the flood event, and that after this flood the warm oceans and cold continents set the stage for the precipitation that later blanketed earth in an Ice Age.

We have seen that the hydroplate theory provides powerful corroboration for the Hopi's own history of their arrival in the Americas over sea from the west, and that it explains the existence of broad land bridges between all the continents for some centuries after the great flood, which enabled migration and then later disappeared beneath rising seas, isolating many people groups.

But such migration would have still taken a long time over land, and does not explain the incredible shared features we have seen in archaeology across the continents that appear to be descended from a single original cultural source. It is possible that migrating peoples carried with them very deeply ingrained legends relating to precession and the heavenly phenomena, and that they preserved these carefully through the millennia, with inevitable alterations between the isolated groups over time, though still distinguishable as belonging to a common source.

However, the technology of the buildings and astronomical angles and symbols that we find on widely separated continents argues for continuing contact by people who were able to cross the oceans after the land bridges were covered by water. In the next chapter, we will examine evidence for such ancient seafarers.

A Theory of Mankind's Ancient Past

In spite of the overwhelming evidence that ancient man had advanced capabilities that surpassed the knowledge of later civilizations by a wide margin, conventional academia rejects or ignores this possibility, and the general public remains completely unaware of the massive amount of data which argues against the conventional paradigm. The conventional framework, with man progressing painstakingly from hunter-gatherer to basic farming and finally to early civilization and onward, reinforces the Darwinian theory of man's origins, and challenges to it cannot be tolerated by those in control of the academic apparatus.

Further, the idea that certain peoples were able to cross the oceans and that cultures currently thought of as largely independent of outside influence (such as the American Indians of North America, the Maori of New Zealand, the Pacific Islanders, and the great civilizations of Central and South America) may have received cultural input of any kind from peoples from the "Old World" (especially if they came from Europe) threatens modern notions of "political correctness" and is attacked as a throwback to racist notions of previous centuries.

But such notions should not be seen as racist, particularly if such contact did occur. In fact, to deny it on the basis of the interests of one group or another could be argued to be the position that is more racist. However, it is important to transcend this kind of name-calling right from the outset and state that if there was in fact a world-wide flood, and if people did survive that flood (as the Hebrew Scriptures and the Christian Bible assert, and as myths and legends in nearly every culture we know of also attest), then all the ethnicities and so-called "races" on the globe today are descendents of those survivors and thus are all related. If that world-encircling catastrophe happened only some thousands of years ago, than we are all related more closely than the proponents of current orthodoxy will even admit. It is important at the outset of this discussion to clearly state that what follows is not motivated at all by a desire to elevate one people-group at the expense of any other.

On the contrary, it seems that Darwin's theory has become such a dominant assumption coloring all of scientific investigation for well over a century that it has created a lower view of mankind than the evidence suggests. Aspects of human capability, such as the existence of energy meridians within the body (the basis for Chinese medicine and the efficacy of acupuncture, among other things), the existence of the energy called *chi* and the ability of some practitioners in China and Tibet and elsewhere to manipulate it to perform feats that Western medicine generally ignores, and many other amazing capabilities of the human mind and body which seem unbelievable from a conventional "scientific" perspective all argue that man is a far more complicated and wonderful being than it has been fashionable to say or even think since the acceptance of Darwinian evolution became the norm.

The same holds for the realization that ancient man, long before the Romans and even before the Egyptians, could traverse the seas, understand the Golden Ratio, calculate the rate of precession to a degree that surpassed even the best efforts of the great Greek astronomers and the calculations of Ptolemy, determine that earth was a globe and measure its dimensions, perhaps even measure the dimensions of the spheres of the moon and the sun, as incredible as that seems. Such facts fly in the face of reigning evolutionary theory, but the evidence suggests that they are facts nonetheless.

Far from elevating one race, ethnicity, or culture above another, this view should make us all amazed at the fact that man is a creation more mysterious and amazing than we have been led to believe by our modern schoolteachers. We can turn to the purveyors of the conventional theory and say, like Theoden to Wormtongue in the novels of J.R.R. Tolkien, "Your leechcraft ere long would have had me walking on all fours like a beast!" (*Two Towers* 158). The accomplishments and capabilities of our ancient ancestors (the ancestors of us all, not of any one group) should be a source of amazement and study, but (like the "spiritual" capabilities of *chi* built into all humans) they are instead ignored or glossed over, when they are not actively ridiculed and scoffed at.

We have seen extensive evidence in myth that some ancient civilization, either a known civilization or more likely a predecessor to known civilizations, understood the celestial mechanics to a high degree, even calculating earth's precession with greater accuracy than was achieved by the Greeks and Romans or for dozens of centuries afterwards. We have seen archaeological evidence from human structures in the "New" World that argues powerfully for connections with those who built structures in the "Old," suggesting the ability to navigate the ocean barriers (and indeed a sophisticated understanding of the heavens and the ability to navigate the ocean wastes are not unrelated skills).

We have seen that one's beliefs about human anthropology are interdependent with one's beliefs about biology and geology, and that the conventional paradigm of human history (which rejects the idea that civilizations could have been more technologically and intellectually advanced in the centuries before known civilizations and known recorded history) is connected to and mutually supporting of the paradigm of uniformitarian geology and the paradigm of Darwinian evolution. We have also seen that Walt Brown's hydroplate theory, which radically upends conventional uniformitarian geology (currently manifested in the plate tectonics theory) threatens the assumptions of Darwinian biology (in that it explains the amazing geologic features found on the earth without requiring hundreds of millions of years). It also helps explain the "anthropological" findings we have discussed so far in this book – the persistent evidence that man's ancient history is far different from what we have been led to believe in the 150 years since Darwinian evolution and uniformitarian geology swept all opposing paradigms from the academic field. Not only does Brown's theory help explain the evidence we find in myth and in archaeology, but that same evidence adds credence to the largely geological and mathematical (in this case, the mathematics of physics) arguments that Brown himself uses to support his theory.

It is only fair at this point to sketch out something of a theory of man's ancient history to connect the dots of the evidence we have examined so far. More study needs to be done in this field before clearer details emerge, but we can begin to flesh out some of the major outlines at least.

First, we must consider that in such a catastrophe as Brown describes, a global flood as described in the Bible and by other cultures around the world, the physical evidence of civilizations previous to such an event would be obliterated. Brown's theory explains that the geologic strata found around the globe are the result of millions of tons of sediments eroded by the escape of high pressure water, dispersed and sorted by the hydrodynamics of liquefaction during the period in which the waters covered the earth. Such an amount of sedimentation would overwhelm any existing monuments of pre-flood mankind. Even a monument as massive as the Great Pyramid itself must have been built after the flood event – it is built upon the sediments left by the flood, for one thing, and it would be ridiculous to argue that a pre-flood structure rode along with the sliding continents and came to rest with its precise astronomical alignments and geodetic alignments all intact.

According to Brown's calculations, various pieces of evidence point to a date for the worldwide flood of approximately 5,000 to 6,000 years ago (based, among other things, on observed mutation rates of mitochondrial DNA, radiocarbon dates of frozen mammoth remains, and the lifespan of comets – which he argues were created during the rupture event by the expulsion of water into space).

Obviously, if no humans survived the flood, then humans must have either been created or arrived on earth since that date. However, if some humans survived the flood, as the Bible and the traditions of many other cultures around the world attest, then it

is possible that those people retained the knowledge and technological understanding of those who lived before the catastrophe.

If, as the Bible declares, human lifespans were significantly longer
before the flood event (which may have somehow altered the
environment to the point that man's lifespan ultimately declined
to the current 120 year limit, which Genesis 6:3 seems to indicate
as a new limit and to associate the initiation of this new limit with
the flood), then it is understandable that correspondingly greater
levels of knowledge and technology might have been achieved
when men lived so long.

However, even without accepting that possibility, it is clear that a
flood date of 5,000 to 6,000 years ago (approximately 3000 to 4000
BC) doesn't leave much time before the advent of the some of the
earliest texts we discussed in previous chapters, such as the Sumerian version of the *Epic of Gilgamesh* (prior to 2000 BC, perhaps as
early as 2100 BC) and the earliest Pyramid Texts (circa 2353 BC
when first inscribed, but possibly first composed centuries earlier
according to some scholars). We have also seen that conventional
scholars date the Great Pyramid to the period of Khufu (2589 BC –
2566 BC). Thus, even if one does not accept the long pre-flood (and,
for a time, post-flood) lifespans described in the book of Genesis,
it appears that the source of ancient knowledge surpassing that
known to conventional history may have come from the people
who lived before the catastrophe.

The reason for this conclusion is that a flood date of 3000 BC or even
4000 BC does not give much time for man to climb from primitive
hunter-gatherer to constructor of the Great Pyramid, which is not
only an engineering marvel but also incorporates precise astronomical alignments indicating high levels of celestial knowledge
as well as measurements that demonstrate an understanding of
the Golden Ratio and even the size of the spherical earth! It is
therefore plausible to believe that the survivors of such a flood
already had advanced knowledge, knowledge that surpassed even
the mathematical achievements of the later Greeks and Romans
(who would discover precession on their own but only at a much
lower level of precision, and some of whom would estimate the
size of the earth, but again not as accurately as their more ancient
predecessors).

In their book *Keeper of Genesis*, Robert Bauval and Graham Hancock examined the ancient Egyptian king lists which are divided into three distinct eras, only the third and most recent of which is accepted as historical by modern conventional academics. The first of these was the time of the *Neteru*, or gods, culminating in the reign of Horus son of Osiris. The second was the era of the Followers of Horus, culminating in the reign of Menes (also known as the Scorpion King or Narmer), followed by the First Dynasty and the rest of the dynastic kings who are widely accepted as historical (192). Bauval and Hancock ask why the last era has been accepted as historical while the previous two are not. Following their line of thinking, we might consider whether perhaps those known in the king lists as "gods" were the generations closely descended from the survivors of the flood, who still possessed the levels of technical capability attained before the catastrophe.

In his writings, including in his book *Ancient Celtic New Zealand*, Martin Doutré has argued that people who predated the people we know as the ancient Egyptians left the region of Egypt and migrated northward, perhaps due to the changing climate that reduced the once-verdant area to a desert, following well-marked trails through Europe as far as England, Ireland, and Scandinavia. As evidence, he provides extensive correlation of features of European megalithic sites – including Stonehenge – and the Great Pyramid. Doutré has provided evidence which argues that measurements at Stonehenge create a half-scale two-dimensional model of the Great Pyramid, complete with the same angles of the sides as in the Great Pyramid itself. The same side angle is also present within the Octagon earthwork found in Newark, Ohio (which the American Indians of the area in the eighteenth century admitted were built by men who were there long before them). This suggests that the pyramid may have come before Stonehenge, and that Stonehenge was a two-dimensional copy made by the exiled descendents of the civilization that built the three-dimensional original!

The conventional theory asserts that the Great Pyramid was built by Khufu (also known by the Greek version Cheops), a pharaoh of the 4th Dynasty who reigned from about 2589 BC to 2566 BC, to serve as a magnificent tomb. However, there are several details which call this commonly accepted piece of conventional wisdom into question.

Most notably, the Great Pyramid is completely devoid of texts and inscriptions, in marked contrast to the tombs of the dynastic pharaohs. The only inscriptions ever found in the Great Pyramid are painted (not carved) hieroglyphics "discovered" by the British Colonel Howard Vyse (1784 – 1853) during his excavation of the pyramid in 1836-1837. These inscriptions, supposed to have been painted by the work-gangs of laborers who hauled the stones during construction, were found in the so-called "relieving chambers," vaulted spaces above the King's Chamber which were supposed in the 1800s to have helped relieve the tremendous weight of the granite above. The lowest of these spaces had already been discovered and opened earlier by Nathaniel Davison, in 1765, and no such graffiti had been found there. It was only when Vyse opened the upper chambers above Davison's that he mysteriously found the only ancient writing in the Great Pyramid.

As Bauval and Hancock point out, Vyse "had both the opportunity and the motive to forge them" (102). They write:

> It is notable that the marks were only discovered in the four 'relieving chambers' opened by Vyse himself, and not in the chamber immediately below these (and immediately above the ceiling of the King's Chamber) which had been opened by a previous explorer, Nathaniel Davison, in 1765. It is also notable that Vyse's diary entry for the day on which he first opened and accessed the lowest of 'his' four chambers (i.e. the one above Davison's Chamber) reports a thorough examination but makes no mention whatsoever of any hieroglyphs prominently daubed on the walls in red paint. On the very next day, however, when Vyse returned to the chamber with witnesses, the hieroglyphs were suddenly there – almost as though they had been painted overnight. 102-103.

The discovery of these red-painted hieroglyphics is pointed to by the conventionalists as proof positive that the Great Pyramid was built in the reign of Khufu. It is remarkable, however, that the strong possibility that they were simply forged by Vyse is dismissed by academia.

Remember that in a previous chapter we noted that conventional academia dismisses an iron plate discovered under the surface of the Great Pyramid's outer edge as being fraudulently planted by Vyse! The presence of iron in a structure built before conventional theories believe that man could make iron is conveniently whisked away under the theory of a hoax by Vyse, but the presence of suspicious painted markings that support conventional theories are declared to be irrefutable evidence that could not possibly be a forgery by Vyse. This is remarkable.

It is noteworthy that the tombs of the later pharaohs, such as that of the 5th Dynasty pharaoh Unas (thought to have reigned from 2375 BC to 2345 BC, less than two hundred years after Khufu), are completely covered with elaborate hieroglyphic inscriptions (deeply carved into the rock). Even more remarkable is the fact that the later Pyramid of Unas is architecturally inferior to the Great Pyramid (and the other two Giza pyramids, which were almost certainly contemporary to the Great Pyramid) in every way, and has long since collapsed into a pile of rubble, in marked contrast to the Great Pyramid. Further, there is no evidence that anyone was ever buried inside the Great Pyramid, or that it in fact was ever intended to be a tomb.

It is completely plausible to argue that the incredible Great Pyramid is the product of a predecessor people to the dynastic Egyptians, and that the later pharaohs then attempted to copy it (without achieving the same level of architectural magnificence) and model their later tombs upon it (perhaps mistakenly believing that it was a tomb). Whether by virtue of their longer pre-flood (and immediate post-flood) lifespans, as described in the Hebrew Scriptures, or some other source of pre-flood civilizational knowledge, this advanced "predecessor civilization" would be the source of the knowledge of the connection between precessional numbers (such as the seventy-two henchmen of Set) and the symbolic "cutting down" of the celestial axis.

If people had an intimate knowledge of the size and shape of the globe and the corresponding celestial mechanics prior to the world-altering flood, and then personally witnessed such a catastrophe, and then observed the changes in the axis and the stars due to the roll of the earth described by Walt Brown in his theory, they would naturally observe the new heavens very carefully afterwards to determine the exact nature of what had taken place. Whether they originally attributed religious significance to such determinations or whether (as is I think more likely) the myths were superlative metaphors for the astronomical processes they observed is immaterial. In fact, what is most likely is that a core cadre or priesthood understood the scientific aspect of this knowledge, preserving it carefully through the succeeding generations in the often violent and brutal centuries as man spread across the earth after the flood, and some people mythologized this knowledge and began to worship the stars, planets and sun while a small few continued to worship their unseen Creator and to forbid such worship of created objects however mighty.

In other words, the "alternative" theorists who perceive that something is very wrong about the conventional academic paradigm and are searching for some catastrophe to explain the destruction of the advanced civilization and who have put forth ideas such as a near-brush of Venus with earth or any number of other theories may have been looking in the wrong place. The answer may be so familiar that they overlooked it – the Biblical flood of Genesis.

Those who subscribed to the idea of a global flood had not given it much thorough scientific examination when they were suddenly confronted by the new theories of men like Lyell and Darwin in the late eighteenth and especially nineteenth centuries. The old explanation, which may well have been correct even though it had not been scientifically examined at all (how could it have been – as the modern understanding of the scientific method was only then being thoroughly worked-out), was overwhelmed and basically swept from the field of battle.

After mopping up the opposition, the devotees of the new theories consolidated their hold on academia to such an extent that scientific examination of evidence for support of the old theories was forbidden (not necessarily by law, but by ridicule, personal attack, and professional ostracism). All new theories would be built upon uniformitarian and Darwinian foundations (new theories such as plate tectonics).

This state of affairs has prevailed for over a century. During that time, few were brave enough to continue looking at the geological record to see if the evidence there was perhaps better explained by a global flood. Walt Brown, however, was one of these few, and his hydroplate theory posits exactly the kind of events that would explain the archaeological and mythological evidence that abounds around the world in quantities too enormous to ignore.

Martin Doutré convincingly argues that the Great Pyramid encoded the crucial measurements of an ancient civilization, measurements which relate to the size of the earth, and their understanding of the mechanics of precession. When this people left the region, moving north along the coast of the Mediterranean and then northward through the Iberian peninsula (and then around the globe), they created new megalithic structures which encoded that same very advanced and cherished knowledge, and which often recalled the dimensions of the Great Pyramid.

For example, he notes that the diameter of Stonehenge was intended to be 378 feet – "exactly half the base length for the Great Pyramid" (164). He further shows that Stonehenge in many ways can be seen as a half-scale, two-dimensional depiction of the Great Pyramid, with a flat-topped "altar" platform at the apex and a face angle of 51.84°, just like that of the monument in Egypt (166).

Since even conventional academicians admit that Stonehenge was begun by 3000 BC, this totally upends the theory that the Great Pyramid was constructed some hundreds of years later by Pharaoh Khufu, unless conventional scholars want to argue that settlers from England influenced the design of Khufu's "tomb"! Far more plausible is the explanation that, for some reason, the possessors of a highly-advanced knowledge of mathematics, astronomy, and architecture – who had lived there for up to 1,000 years after the flood event – chose to leave the region of Egypt (perhaps because of severe weather changes, or because of conflict with other people entering the region), and that they encoded their most important information wherever they ended up. They influenced those who we know as the dynastic Egyptians, who revered them and imitated them, and who tell of these mysterious predecessors in the "king lists" that scholars today reject as legend.

Note that the thesis of Jane Sellers in *Death of Gods in Ancient Egypt* may in fact be used as an argument to support this conclusion as well, although she herself does not subscribe to the idea of a highly advanced ancient civilization. Her thesis places the origin of the myth of Osiris and Seth in the distant *pre-dynastic* era. Based on the other evidence we have examined in this book, it seems clear that the pre-dynastic civilization was more advanced than the later dynastic period, rather than less advanced, even if Jane Sellers herself would not agree with that conclusion.

Who were these ancient possessors of such remarkable skills and knowledge? Where did they go? Some certainly remained in Egypt. Based on the evidence, those that left may well have traveled east and north through the Holy Land, where some remained to become the Canaanites and Phoenicians. There is extensive evidence that the ancient Phoenicians had ships that could and did travel regularly to the Americas and even as far as Asia and Australia.

The fact that the at least some of the pyramids in Central America also appear to have scaled proportional relationships to the Great Pyramid, and the fact that the Aztecs of Mexico told the first Spaniards who asked that they themselves did not build them but that they were built by an earlier people whom the Aztecs called the Toltecs, indicates that members of this ancient civilization crossed the oceans soon after their expulsion from Egypt and settled in the New World as well.

Further north from Canaan, the descendents of this ancient civilization may have continued to Anatolia, an ancient Greek name for the western portion of modern Turkey. The Anatolians were feared and aggressive sea people who raided all along the shores of the Mediterranean.

We recall that in the discussion of the Mithraic mysteries, Plutarch identified the origins of the cult of Mithras with the Anatolian pirates from the region of Cilicia, who may have been the descendents of these same more ancient Anatolians and whose cultus might have retained some of those peoples' astronomical knowledge and symbology.

Professor Barry Fell (1917 – 1994), author of *America BC: Ancient Settlers in the New World,* also notes that the later Libyans were very likely the descendents of the same Anatolian sea people who attempted to invade Egypt and then later settled in Libya around 1400 BC. Like the later pirates from Anatolia described by Plutarch, these Libyan descendents of the Anatolians were skilled sailors, often "employed by the Pharaohs in the Egyptian fleet" (176). Based on Libyan inscriptions discovered on stones in both North America and South America, Fell believes these Libyans made extensive excursions across the oceans as early as 1000 BC and certainly prior to 500 BC (177).

Based on linguistic analysis of the extensive parallels in language, Fell also believes that the ancient Libyans influenced southwest American Indians such as the Zuni people (or A:Shiwi), as well as the Polynesians of the Pacific islands (176).

Those descendents of the ancient civilization in question who did not stop in Canaan or Anatolia may have continued further, progressing along the northern part of the Mediterranean into Greece and on to Spain, being there already when ancient Phoenicians founded the city of Tartessos or Tarshish (a name which is identified with one of the sons of Javan son of Japheth in Genesis 10:4). The recent discovery in 2011 of a city answering to some of Plato's descriptions of Atlantis in southern Spain appears to strengthen this theory.

The Hebrews also appear to have some ancient connection with the people who anciently settled in what is modern Spain. Many scholars have pointed out that the ancient name for Spain, *Iberia,* is derived from the same root as the word Hebrew, which itself denotes the descendents of Heber or Eber, named in the book of Genesis as the son of Arphaxad, the son of Shem, one of the three sons of Noah (Genesis 10:23-25).

258 The Mathisen Corollary

Martin Doutré argues that the Druidic religion of Europe was
essentially Hebrew, that it valued knowledge such as the motions
of the stars and planets but that this knowledge was incorrectly
misinterpreted as worship of the sun and stars, and that their
astronomical circles and pentagrams were violently suppressed
by the later Church of Rome as diabolical, which was ignorant of
their true purpose. He writes:

> The Druids, described by Caesar as a *'Disciplina,'* were
> highly enlightened, with some categories of qualification
> within their schools requiring 20 years of intense training
> before completion. Their Schools of Learning offered many
> separate branches of specialised tutorials, in the arts and
> sciences.
>
> [. . .]
>
> The Druids were a quasi-religious Order in the sense that
> they venerated a "Creator" and lived in harmony with their
> surroundings. Their Order, like the Levitical Order of the
> Hebrews, was set apart to work for, support and bless, the
> general population. 228.

He also points out that nineteenth-century scholars, unencum-
bered by the political correctness that marks scholarship today,
often noticed the same strong connections between Druidic and
Hebrew theology. Doutré notes on page 227 of his *Ancient Celtic
New Zealand* that Cassell's *Illustrated History of England* (1873)
declares "The Druidic Rites and Ceremonies in Britain were almost
identical with the Mosaic Ritual" and that Professor Charles
Hubert in 1825 asserted that "So near is the resemblance between
the Druidic religion of Britain and the Patriarchal religion of the
Hebrews, that we hesitate not to pronounce their origins the same."

As powerful confirmation of this theory, Doutré points to the
passage found in II Chronicles 4:1 of the Hebrew Scriptures
(chapter and verse as divided in the AV Old Testament of the
Bible), describing the dimensions of the altar in the Temple built
by Solomon. There, the altar is described as being "twenty cubits
the length thereof, and twenty cubits the breadth thereof, and
ten cubits the height thereof." Using a standard cubit of eighteen

inches (as opposed to a royal cubit of twenty-one inches), Doutré points out that the four sides of 20 cubits each have a perimeter of 1,440 inches (360 inches each side or 20 cubits of eighteen inches each). When multiplied by the height of 180 inches (or 10 cubits of eighteen inches each), the dimensions will yield 259,200 (Doutré 136). This number, of course, is a precessional number, and equals ten times the number of years in the full cycle of precession, if the precession is calculated to alter the heavens by one full degree every 72 years (an approximation, as we have already discussed, but a much closer one than calculated by any known scholars of antiquity).

Doutré cites many other coded numbers in the oldest Hebrew Scriptures. We could point out as an example the dimensions of Solomon's porch, described in II Chronicles 3:4, which – while not specifically cited by Martin Doutré – is given to be twenty cubits in breadth and one hundred twenty in height. Using eighteen-inch cubits, this yields an area of 43,200 square inches for the area cited in that verse. In the preceding verse, the building itself is described as having a length of threescore cubits and a breadth of twenty cubits, which yields an area of 21,600 square inches. As we have already seen, these are important precessional numbers. It is certainly possible that the ancient Hebrews incorporated these dimensions into their most sacred structures without implying acceptance of a worship of the heavens, which they obviously strongly condemned (see passages in the same book of II Chronicles, such as II Chronicles 33:5 and II Chronicles 34:3-5, as well as Job 31:26-28 and indeed almost any of the other ancient Hebrew holy books). But it is also clear that the temptation to devolve into worship of the sun and the stars was strong and that a running theme of the sacred Hebrew writings is the tension between those who succumbed to that temptation and those who did not.

Doutré argues that the builders of the Giza Pyramids emigrated from Egypt to either influence or to actually become the Phoenicians and the Celtic peoples of Western Europe and the British Isles (and possibly the early Nordic people as well, with whom they share notable physical characteristics including the very rare red-hair gene and a history of seafaring). They were then responsible for building Stonehenge, the megalithic sites of Spain, France and Ireland, and with their superior naval capabilities traversed the globe, creating megalithic sites as far away as the Americas, Asia including China and Korea (although it is possible that these regions were influenced by cultural contact over land from the direction of Egypt by way of India over the centuries rather than by sea), and even New Zealand. These sites were meant to serve as versions of the Great Pyramid and the Giza Pyramids for a civilization in exile. Ocean-crossing capabilities of this people, particularly the Phoenicians and the dwellers at Tarshish (or Tartessos, as the Greeks called it) in Spain are described in the Hebrew Scriptures, as well as in the later writings of Herodotus and Hesiod (and later Julius Caesar), indicating that these ocean-crossing voyages went on for many centuries before the Phoenicians were finally destroyed by the Romans. The evidence of their ongoing trade can be found in the mummies of the dynastic Egyptians, for example, which have been found to contain traces of New World plants such as tobacco and cocaine, although of course conventional scholars deny such findings.

Extensive and dramatic evidence supporting this broad theory of ancient history was documented by professor Barry Fell, most notably in his book *America BC* (1976), which contains photographs and discussion of hundreds of New World inscriptions carved in stone using very old forms of Iberian and Celtic and Libyan writing. Many of these inscriptions were found by early seventeenth-century colonists (including Cotton Mather in 1712) at a time when they could not have possibly forged them, since the alphabets used were only deciphered much later by scholars, and can now be translated and understood. While conventional academia has taken pains to dismiss Fell as a fraud, it is impossible to dispute the existence of these inscriptions, for they are too numerous to deny, as well as the existence of the stone structures known as "root cellars" (which the first American settlers in the colonial days found and described in detail) that he maintained were evidence of ancient civilizations – and which in fact bear striking resemblance to the extensive megalithic sites in the Boyne River region of Ireland, albeit on a smaller scale.

The inscriptions found in America described by Barry Fell in *America BC* include at least four different ancient writing systems, which encode inscriptions written in several different ancient languages. Sometimes inscriptions from the same language will be written in two different alphabetical systems. The first of the four different writing systems and languages is a form of Ogham that is listed in a medieval tract, *Book of Ballymote*, written at some point between the 1370s to 1390s AD in Gaelic by Irish monks. Actual examples of some of these forms of Ogham listed in the *Book of Ballymote* had never been found before in Europe and were assumed by scholars to have been a fanciful invention of the monks who wrote the tract. Because these forms were not known from Irish specimens, European scholars dismissed them as monkish inventions, unaware that Americans had discovered examples of these forms in the US. The Americans, for their part, were unaware of the medieval text that could unlock the mystery. Later, thanks to the efforts of Fell and other epigraphers, the puzzle would be unlocked.

More curious still, the inscriptions in America that use this ancient Ogham alphabet are written in various languages, one of them being a form of Phoenician used in ancient times by Phoenicians of the Iberian Peninsula in the period 800 BC to 600 BC (Fell 45-57). Another language used is Libyan, an ancient language very closely related to ancient Egyptian and probably descended from it, with similarities to modern Coptic (Fell 64, 178). Fell describes it as "basically Egyptian combined with Anatolian roots introduced by the Sea Peoples who invaded Libya" around 1400 BC (174), and he traces fascinating connections to the languages of the Zuni of New Mexico and also to the people of the Pacific Islands, where other inscriptions using the same writing are found.

The second of the writing systems employed on inscriptions found so far in the Americas is a form of Iberian script closely related to other forms of Phoenician writing. This script contains letters as opposed to the "grooves" or lines of Ogham, and because some of our own modern alphabet is descended from Phoenician (by way of Greece and then Rome), some of these letters are similar to letters we still use today. The language used in the inscriptions that employ this Iberian alphabet is the same dialect of Phoenician used in the inscriptions that use the Ogham writing system, as if the Iberian sailors who visited this continent could use either script in order to communicate with people familiar with only one or the other (Fell, 64). Fell relates that lead lamina found in Spain also employ the Greek alphabet while still using the Phoenician language (104).

A third form of writing system used for inscriptions in the New World is the Libyan alphabet. As we might expect, the same Libyan language found in some American Ogham inscriptions is also employed in inscriptions that use the Libyan letters, which Fell describes as "like that of the Phoenicians, alphabetic but using only consonants" (174).

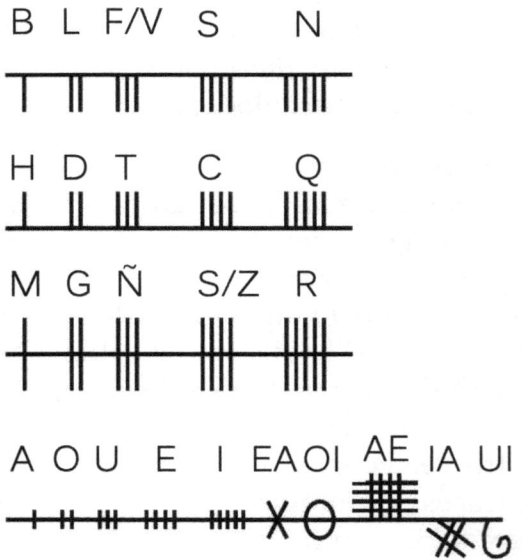

An Ogham stone in Ireland. This writing system was used by the Goidelic or Q-Celts in Britain, and it survived longest in Ireland. The diagram to the right shows a common horizontal Ogham; when vertical the marks that would be below a horizontal stem-line are then made to the right of the vertical stem-line. When carved using a corner or edge on a stone as the stem-line it is known as "edge Ogham." Dr. Barry Fell observed that many variants are possible; in Iberian Ogham there are fewer consonants and vowels are omitted. This is the version that is found in the Americas and, Fell believes, the oldest form. Alphabet from *America BC*, 47.
image: Wikimedia commons.

A fourth form documented in the book is hieratic, a less formal version of Egyptian hieroglyphics than the formal hieroglyphics familiar from the grand temple monuments, but still based on the formal hieroglyphics, although in a simplified and more easily-written format. Inscriptions using this writing system are in the Egyptian language.

Fell documents numerous examples of inscriptions using the cyrptic Ogham (or Ogam) alphabet, which was a Celtic alphabet-ical system using carved lines on either side of a stem-line, which could be either horizontal or vertical, to indicate letters. We have previously mentioned some of the observations on Ogham in previous chapters, such as the observation by Robert Graves (1895 – 1985) that the entire alphabet contains 72 strokes. The mytho-poetic traditions associated with this alphabetical system are extensively dealt with in *The White Goddess: A Historical Grammar of Poetic Myth* (1948), by Graves. While *The White Goddess* covers an enormous amount of ground, much of it controversial and (unsurprisingly) attacked by conventional scholars, the pertinent points to this discussion are the connections he found between the Ogham alphabet and the cult of the zodiac and the heavenly motions, the connections between its letters and European trees (and hence a connection to one aspect of Druidism), and his explo-ration of the deep connections between pagan religions found throughout history (and, as we have seen, throughout the world).

As Graves correctly points out, Ogham is most commonly found in the British Isles, and especially Ireland. As Barry Fell explains in *America BC*, prior to discoveries in the 1960s and 1970s, the consensus was that "Only the Gaelic Celts of Britain seem to have been acquainted with the Ogam system of writing. The Gauls of France and the P-Celts of ancient Britain employed Greek or Latin letters, as their coins attest" (63).

By "P-Celts" are indicated one of the two large divisions of Celtic tongues identified by preeminent Celtic scholar Sir John Rhys (1840 – 1915), who noted that one group, the "Q-Celts," contained words that used a "c" or "q" sound for corresponding words in which the "P-Celts" use a "p" or "b" sound. For example, the Gaelic and Irish word for "five," *coig* or *cuig* is descended from

Q-Celtic, while the corresponding Welsh word *pump* is descended from P-Celtic. Interestingly, the Q-Celtic words for "five" correspond to the Latin *quinque* which also uses a "q" sound, while the P-Celtic word for "five" corresponds to Greek *pente*, which also uses a "p" sound (Fell 40-43). Tradition and some scholarship suggests that these Ogham-using Q-Celts were the ancient Gaels or Goidels, who had come to the British Isles from somewhere else (as we shall see, probably from Spain) and were later driven out of England and Wales by the P-Celt Gauls from France, who "contented themselves with Greek or Latin letters, as we see in Gaulish inscriptions in France" Fell says (43).

Concerning the origin of the Q-Celts, Fell notes that:

> One of the ancient names of Ireland is *Ibheriu*, pronounced as *Iveriu*, a fact that suggests that the word is derived from a still-earlier pronunciation, *Iberiu*. Now this is very interesting, for the Gaelic histories assert that the ancestors of the Gaels came to Ireland from Iberia, the old name of Spain. Could *Iberiu* be the same as *Iberia*, the name of the older homeland having been transferred to the younger?

Plaster cast of the Grave Creek tablet (Moundsville tablet), found in West Virginia in 1838. At the time, Iberian writing could not even be deciphered yet. It was generally attributed to American Indians throughout the 1800s. The original has now disappeared.

Many people, including some linguists, think this may well be the case. 43.

We have already seen the etymological similarity between the word *Hebrew* and *Iberia*. If Fell's theory is correct, we can link it to the arguments of Martin Doutré and assert that the builders of the Giza Pyramids left Egypt somewhere prior to 3000 BC, moving northward through Canaan and either influencing or becoming the ancient Phoenicians / Hebrews, continuing to emigrate to Anatolia, Greece, southern Europe and Spain in particular (where they founded Tartessos or Tarshish), and on into the British Isles, where they settled in Ireland and England. Those who settled in Ireland continued to use Ogham after their cousins in Britain, Wales and Cornwall were overwhelmed by the influx of the P-Celt Gauls from France.

Fell went in search of archaeological evidence to support his theory of a connection between Spain and the Ogham-using Q-Celts of Ireland, and *America BC* chronicles the discoveries in Spain and Portugal of hitherto-unknown Ogham inscriptions from Lerida, Spain and from the Duoro River valley in northern Portugal (63 – 74). More remarkable still, these versions of Ogham appeared to be older than those found in the British Isles, based upon a listing of alphabets preserved in the *Book of Ballymote*, and on the remarkable fact that some of the Ogham writing of the Iberian Peninsula omits vowels, much as the Phoenicians and Hebrew writing systems do (64). Prior to this discovery (and even today, over thirty years later), scholars maintained that the Ogham alphabet found on line 16 in the *Book of Ballymote* was the only authentic alphabet, and that the apparently earlier variants found on lines 1-15 of the Ogham Tract of that medieval book were simply imaginary creations made by monks to pass the time.

The discovery of older examples of Ogham on the continent should cause those who argue that the alphabet was not developed until the first century AD to revise their theories. However, although Fell was optimistic that the discoveries he published in the 1960s and 1970s would cause great changes in academic thought, the stirrings of interest among university professors that he described back in 1976 have been buried under the weight of four decades of academic orthodoxy.

It is this earlier form of Ogham which appears in inscriptions and monuments in the New World. Fell documents dozens of these inscriptions in *America BC*. For his efforts, he is today slandered as a fraud and a racist by defenders of the academic dogma that regards any suggestion of ancient Old World contact with the New to be anathema.

In his book, Fell documents numerous examples of Ogham found in North America. He explains:

> About three thousand years ago bands of roving Celtic mariners crossed the North Atlantic to discover, and then to colonize, North America. They came from Spain and Portugal, by way of the Canary Islands, sailing the trade winds as Columbus also was to do long afterward. The advantage of this route is that the winds favor a crossing from east to west, but for Celts accustomed to a temperate climate it had the one drawback that it led them to the tropical West Indies, no place for northerners. So although their landfall lay in the Caribbean, it was on the rocky coasts and mountainous hinterlands of New England that most of these wanderers finally landed, there to establish a new European kingdom which they called Iargalon, "Land Beyond the Sunset." They built villages and temples, raised Druids' circles and buried their dead in marked graves. They were still there in the time of Julius Caesar, as is attested by an inscribed monolith on which the date of celebration of the great Celtic festival of Beltane (Mayday) is given in Roman numerals appropriate to the reformed Julian calendar introduced in 46 BC. 6

Note that Beltane or Mayday is a "cross-quarter day" – midway between the March equinox and the June solstice, and was typically celebrated on the 6th of May as the beginning of summer (later, for simplicity, most cross-quarter days were moved to the first of the month, so that the old celebration of the 6th of May moved to the 1st of May, or May Day).

Fell provides extensive descriptions and photographic evidence of Ogham texts carved into stones found in Vermont, New Hampshire, Connecticut, New York, Texas, Oklahoma, Arkansas, and the Caribbean. He provides translations of many of these, usually read from right to left. Among them:

> To Baal Son of Iabagug [or Habakuk]. 54

> Stone of Bel [probably Stone of Beltane, Fell remarks]. 22

> Y-G-H-N [a proper name, Yoghan – "In modern spelling this would be rendered as Eoghan," Fell explains]. 59

Fell explains that in Europe, "when Christianity came to the Celts the priests caused all the ancient pagan inscriptions to be erased, replaced by Christian Ogam, or left blank, while all the offending fertility paraphernalia were totally destroyed. Not so in America. Here Christianity never came to the Celts, their old pagan inscriptions remain intact [. . .]" 7.

But Ogham is not the only ancient alphabet to be found in the Americas. Fell continues with his timeline:

> In the wake of the Celtic pioneers came the Phoenician traders of Spain, men from Cadiz who spoke the Punic tongue, but wrote it in the peculiar style of lettering known as Iberian script. Although some of these traders seem to have settled only on the coast, and then only temporarily, leaving a few engraved stones to mark their visits or record their claims of territorial annexation, other Phoenicians remained here and, together with Egyptian miners, became part of the Wabanaki tribe of New England. Further south, Basque sailors came to Pennsylvania and established a temporary settlement there, leaving however no substantial monuments other than grave markers bearing their names. Further south still, Libyan and Egyptian mariners entered the Mississippi from the Gulf of Mexico, penetrating inland to Iowa and the Dakotas, and westward along the Arkansas and Cimarron Rivers, to leave behind inscribed records of their presence. Norse and Basque visitors reached the Gulf of St. Lawrence, introducing various mariners' terms into the language of the northern Algorquian Indians. Descendants of these visitors are also to be found apparently among the Amerindian tribes, several of which employ dialects derived in part from the ancient tongues of Phoenicia and North Africa. 7.

Fell believes based primarily upon the languages and types of writing found that deliberate voyages began from the Old World to the New circa 1000 BC, and that the first Ogham and Iberian Punic inscriptions date from around 800 BC.

Fell provides examples of Iberian runes found in American sites located in West Virginia, Pennsylvania, Massachusetts, New Hampshire, and central Vermont. A version of this alphabet was used by the men of Tarshish, and it was not deciphered until the late 1960s. Clear parallels between inscriptions found in the Americas in the early 1800s and those found in Spain are apparent in the image at left.

The Phoenician characteristics of the writing on this and other tablets are quite evident, and can easily be compared to any of the hundreds of examples that exist from the Mediterranean, such as the well-known Nora Stele of Sardinia. To continue to say that Fell's identification of these inscriptions from the New World is "controversial" or in any way "strained" or "speculative" is ridicu-

Comparison of the writing system on the Grave Creek Stone (discovered in 1838 in West Virginia, before Iberian script could be deciphered) with the Nora Stele of Sardinia, containing Iberian writing from the 9th century BC. The reader can see the clear similarities between many letters. Such comparisons are often called "controversial," but they should be called "obvious."

lous. Fell did not "claim" that the Moundsville Tablet and other artifacts from North America contain Iberian Punic (Phoenician) writing: they clearly do contain such writing! To continue to deny this or to claim that some fraudster in West Virginia in 1838 created a tiny stone measuring an inch and a half by an inch and seven-eighths and carved into it some lines that turned out to match ancient Phoenician is also ludicrous.

The Phoenician city in Spain of Tarshish was known throughout the ancient world for the seagoing prowess of its sailors. Iberian inscriptions in Bristol, Rhode Island and Monhegan Island off the coast of Maine testify to the ongoing visits of seafarers from Tarshish to North America (mentioning it by name).

Fell also identifies the markings on another North American find, the Bourne Stone, as Phoenician Iberian. The Bourne Stone is a 300-pound block of pink granite found in Massachusetts in 1680. According to historians, it was used as the doorstep of an Indian Mission church and was turned over with the inscription down because the Native Americans refused to step on the ancient markings. Later, the same stone was used in the foundation of a house, where it was noticed in 1800 and recognized as an artifact due to its markings.

The fact that these inscriptions were mentioned prior to 1800 makes it more difficult to laugh them off as forgeries. Instead, conventional academia seems to have implemented a policy of ignoring this artifact (how many American schoolchildren, or even college graduates, are even aware of its existence?) and – for those who do learn of it – stating that the inscriptions are difficult to interpret and that a host of different languages have been suggested, none of them conclusive. The web page of the Bourne Historical Society, which has the stone on display, shows no images of the stone and declares that "many attempts over the ensuing years have still not conclusively revealed the origin of the carver or the meanings of these markings; the last investigation was conducted in 2003 in St. Paul, MN" (http://www.bournehistoricalsociety.org/bourne-stone.html, accessed 03/25/2011).

No mention of the possibility of a Phoenician origin of the carvings is made.

However, as with the Grave Creek Stone, the parallels to known Phoenician-inscribed artifacts from the Old World are quite striking and hard to deny. The inscription consists of two lines, with the letters of the upper line much larger than those of the lower. Fell compares the individual letters side-by-side with Iberian letters from southern Spain; many are identical, some are mirror images, and the "Q" (which is quite prominent in the upper line of the Bourne inscription) is a simplification of the corresponding Phoenician symbol, which looks like a figure-eight on its side (our mathematical symbol for "infinity") with a vertical line through it, which became our letter "Q" (Fell 160-161).

More remarkable still are some of the other artifacts containing ancient writing found in North America during previous centuries and containing images which we can now (following the arguments in *Hamlet's Mill* and the further examination we have made in this book) understand as having to do with the celestial mechanics of precession. Most intriguing of these, perhaps, is the Davenport stele, which depicts a scene we have encountered before, in the imagery of Set and Horus "turning the drill" in ancient Egypt, as well as in Hindu imagery as the "churning of the milky sea" and even in the Mayan Codices preserved from destruction after the Spanish conquistadors and priests destroyed most of the records of Central American civilizations.

This stele, which has been dismissed as a fraud and barely examined by scholars for over a hundred years, contains writing which Fell identifies as using three different alphabets (much like the famed Rosetta Stone). The first is Libyan, the second is Punic-Iberian, and the third is Egyptian Hieratic.

An outsider such as Sherlock Holmes or the team from Scooby Doo, when confronted with this amazing clue and the consensus opinion of the "authorities" that this clue is a fraud, might immediately start to ask some probing questions. Why, for example, would a forger in the late 1800s bother to inscribe their hoax with three different alphabets and three different languages? Is it just an amazing coincidence that they selected three different cultures for their forgeries, all of which left other inscriptions around the North American continent, even though the existence of these other Libyan, Phoenician and Egyptian Hieratic inscriptions was hardly common knowledge (nor is it today)? Isn't it a stretch to believe that these forgers were so meticulous that they actually selected the correct versions of these arcane languages (aspects of which were not even deciphered until the twentieth century) to

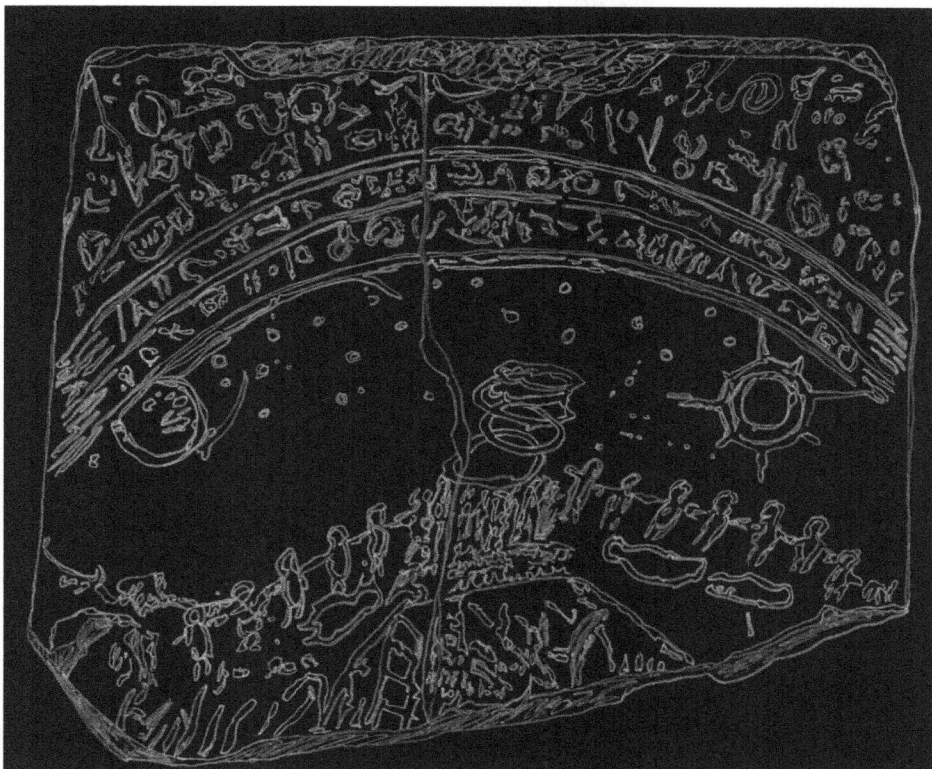

One of three stones found in 1874 near Davenport, Iowa, known as the Davenport Stele or the Davenport Tablets. These have been confidently labeled as fraudulent, and undergone very little research in the twentieth or twentyfirst centuries. If they are forgeries, it is absolutely amazing that the forger knew enough to depict the scene we have discussed from the throne of Sesostris I and from the Mayan Codex Tro-Cortesianus and from the Hindu Vedas known as the "uniting of Upper and Lower Egypt" and the "churning of the ocean of milk." The importance of this scene was not widely known before the controversial thesis of de Santillana and von Dechend published in 1969. Critics who have declared it a fraud often incorrectly identify the central pillar as "smoke rising from a fire," demonstrating the ongoing ignorance of the importance of this scene and unwittingly illustrating how unlikely it is that a forger would choose to depict it.

match the historical periods at which Fell believes (based on the evidence) the Phoenicians and Libyans were exploring the New World?

More puzzling still is the image itself, which would seem a very curious choice to be carved by nineteenth-century forgers. The scene depicts a trapezoidal mound upon which are some sort of vertical lines suggestive of a pyre of sticks, with spiraling circles rising above them. Human figures are stretched out to either side of this pyre, their arms apparently linked, and other human figures appear to be lying on the ground at their feet, as if they are perhaps dead or even bundled like mummies. On the right side of the scene is a brilliant sun, and on the left a disc that has been taken to be a moon (but which Fell argues contains Egyptian hieroglyphs identifying it as a mirror to reflect the rising sun onto the center of the scene), with smaller circles that may be indicative of stars in between.

Critics have argued that the supine forms on the ground are sacrifices, and that if so it would make sense that they would be

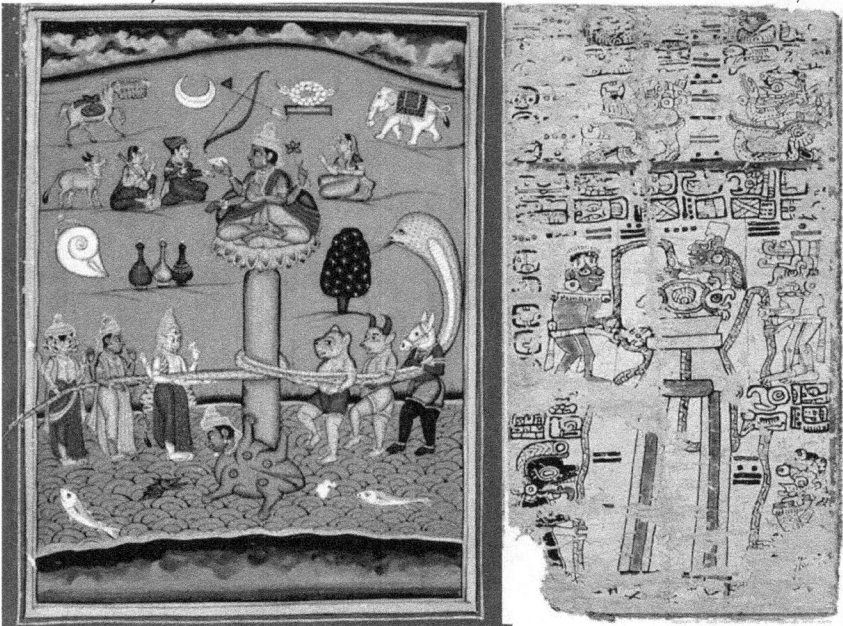

Scenes from the ancient Hindu "churning of the ocean of milk" and the Mayan Codex Tro-Cortesianus. Compare to the scene in the Davenport Stele.

depicted in the fire rather than on the ground at a distance from it – thus it must be a forgery! This is a ridiculous line of argument, but it was advanced by one of the early "debunkers" of the stele, Cyrus Thomas (1825 – 1910), who originally hoped the stele would support his "mound-builder" theories but concluded it was a hoax. His arguments are sometimes brought up by more recent critics as evidence that even contemporaries to the forgery could spot it as a fake.

In fact, it is debatable whether the linked rings rising from the trapezoidal mountain in the center of the scene represent the smoke of a sacrificial pyre at all. As Fell himself points out, the central image is very reminiscent of the Djed pillar, "made of bundles of reeds encircled at the top by rings, [and which] represents the backbone of Osiris" (266).

It is incredible that the supposed forgers would have thought to depict a Djed column on a tablet with Libyan, Punic-Iberian, and Hieratic writing over it. However, debunkers have posited that the forgers were learned scholars who wished to embarrass and discredit the original finder of the artifact, the Reverend Jacob Gass, an immigrant from Switzerland. Even if we grant the hypothesis of very learned forgers who knew of Egyptian Djed columns and decided to forge one on the tablet, why would they have done so in such an unorthodox manner, rather than drawing a traditional stylized Djed column of which numerous examples survive from ancient Egyptian art? The depiction in the Davenport stele is almost a "shorthand" for a Djed column, done by someone who was so familiar with them and whose intended audience was so familiar with them that he would not need to be very deliberate to convey his intended message.

More remarkable still is the addition of lines of humanoid figures on either side of the Djed column, arrayed as if holding hands or perhaps playing tug-of-war with a rope. Fell interprets these as "worshippers hauling ropes to raise the Djed column" (266), and this may well be the intent of the engraver, but other Djed depictions from Egypt do not typically feature such worshippers hauling ropes. Instead, this scene immediately brings to mind the churning of the milky ocean from the Hindu Puranas, and the two beings pulling a snake-like rope in the Maya Codex Tro-Cortesianus, all of which we have argued are precessional metaphors relating to the turning of the mighty mill of the sky and the understanding that the central axis of heaven was balanced between the forces of order and chaos, Horus and Set, turning out a pattern of great regularity which was nevertheless skewed and slipping, creating the precession of the equinoxes over the long millennia.

Again, an investigator would have to ask whether the supposed nineteenth-century forgers decided to use an unfamiliar but clearly-recognizable way of depicting the Djed column, and then attach two sets of figures pulling something attached to it, and if they did so, what was their inspiration? Did they understand ancient connections between the forerunners of Hinduism, the Egyptians, and the Maya, and if so, who were these brilliant forgers? Far more likely an explanation is that the Davenport stele is not the work of forgers at all, but an extremely important American artifact and confirming clue that ancient history did not happen exactly the way the keepers of the conventional storyline would have everyone believe.

But, even if the Davenport stele is a fraud (and it would be remarkable, given the facts discussed above), and even if Fell somehow manufactured or imported all the examples of Ogham that he claimed to find in the New World (a ridiculous assertion, but one we will make for the sake of argument), nobody in his right mind would suggest that Fell and other co-conspirators also forged the so-called "root cellars" discovered by some of the earliest American colonists in the 1600s and 1700s, some of which still survive in New England to this day.

These bear a striking and significant resemblance to the corbelled mound-passages of the Burgoyne Valley in Ireland which are described in detail in Martin Brennan's *The Stars and the Stones: Ancient Art and Astronomy in Ireland* (1983). Fell could not have known of this work when he published *America BC* and its discussion of the "root cellars," because Brennan had not yet written it.

In fact, Brennan and his colleagues were pursuing a novel theory that the megalithic passages in the mounds of Ireland were not burial shafts at all but served an astronomical purpose, allowing precise measurements of the sun's beams on significant days, particularly the equinoxes, solstices, and "cross-quarter days" in between the equinoxes and solstices. This theory was in no way widely accepted and it caused something of a bombshell when Brennan and his friends, after diligent and tedious work (often in the cold, damp, early morning hours of many solstices and equinoxes) provided substantial proof that their theory was correct – a bombshell that did not hit the news until March of 1980.

Throughout New England, early settlers in the 1600s and 1700s began to encounter structures composed of stones, which they took to be the work of ancient American Indians. Some of the stones of these structures, which took the form of underground chambers and hallways around eight to ten feet wide and high and sometimes up to thirty feet in length, weighed several hundred pounds. When discovered, they were sometimes used to store crops (hence early colonists called them "root cellars"), and even more often they were used as sources of stone for bridges and buildings by the settlers.

Barry Fell attributes the lack of scholarly interest in these structures, some of which contained carvings, to the decisive split between England and the colonists in 1776, after which the British scholars who began studying discoveries of similar inscriptions and structures in Ireland and England were not able to link them to the megaliths in America, and any Americans who thought there might be more to the "root cellars" did not have access to the megaliths in England and Ireland that would have enabled them to put the two together (Fell 10-12).

In spite of depredations by those who hauled away the stones over the past three centuries, several of these stone structures still survive, including a complex near North Salem, New Hampshire known as Mystery Hill (first discovered in 1823); the Upton Cave in Upton, Massachusetts; a complex at Gungywamp in Groton, Connecticut; and several chambers in New York including at Druid Hill, Mead's Corners, and Putnam Valley.

These chambers cannot be easily dismissed as forgeries or hoaxes, the way tablets such as the Davenport stele can be (even though such dismissals, as we have seen, appear to be foolish rejection of important clues to the mystery of man's ancient past). For one thing, they are too large to attribute to forgers, since moving slabs weighing a ton is not easy to do, and doing so surreptitiously would be out of the question. Second, they bear all the markings of great age, including heavy growth of lichens and even mature trees growing above them. Third, they were plainly attested to by early settlers, who believed they were the work of people there before them. Nevertheless, some modern defenders of the conventional storyline argue that these stone chambers were constructed by colonists in the 1600s and 1700s to store their root crops, even when the chambers are clearly shown to contain sophisticated alignments with celestial events such as the equinoxes! Would early Puritan settlers take the trouble to align their potato storage sheds with the equinoxes? The entire suggestion is absurd.

Moreover, the parallels between these existing "root cellars" and the megalithic mounds of Ireland (such as at Dowth, Knowth and Newgrange in the Boyne Valley, in the Loughcrew Mountains, and at Fourknocks, Knockmany, Sess Kilgreen, and Carrowmore, among others), are striking. Long believed to be burial chambers, the megalithic passages of Ireland were decisively shown to be sophisticated structures for determining important days of the year such as the equinoxes, solstices, and cross-quarter days by Martin Brennan and his colleagues in the late 1970s and early 1980s.

There had always been folk traditions associating these Irish chambers with strange phenomena of the sun and moon, and the stones themselves contain extensive engravings of stars, suns, and spirals suggestive of an astronomical connection, but modern scholars prior to Brennan's work had dismissed such ideas. "In this century," Brennan explains, "solar imagery was almost completely ignored, because it could be too closely associated with astronomical alignments, which did not fit in with preconceived notions of prehistoric cultural development" (35). The reason conventional scholars were so adamant in denying any sophisticated astronomical purpose to the ancient megalithic mounds of

Ireland is their great age: they are acknowledged to be over 5,000 years old (built before 3000 BC).

The parallels between the structures in Ireland such as those at Newgrange and Loughcrew and the "root cellars" of North America include the use of large stones, the precise alignment with solstices and equinoxes, and the occasional use of a unique roof architecture called "corbelling." A corbel is an architectural feature in which a portion of a stone is left to jut out of a wall in order to support a weight above, and it is still used today. In ancient architecture, successive layers of stones would be corbelled to create a vaulted roof, as shown below.

Some of the stone structures in New Hampshire (but by no means all of them) feature corbelled arches in their main chamber, just as some of the megalithic structures in Ireland (but by no means all of them) also feature the same type of corbelling.

Like their counterparts in Ireland described by Brennan, the surviving stone structures of New England are almost invariably aligned with the sunrise point on the equinox (due east, the direction at which the sun always rises on the equinox worldwide), the summer solstice, or the winter solstice.

Martin Brennan in *The Stars and the Stones* explains why megalithic hallways facing towards the sunrise provide a more precise and sophisticated indicator of specific calendar days than a gnomon or obelisk casting a shadow on the ground:

> The passage and chamber of a megalithic mound is a far more complex, precise and specialized instrument than a gnomon is. Instead of simply casting a shadow, the passage narrows a beam of light as it is projected into the chamber. The passage is aligned to particular points on or near the horizon. When the sun occupies these points at special times of the year, light passes without interruption along the whole length of the passage to illuminate the backstone of the chamber. This type of sundialling is called sunbeam or light-beam dialing, and it represents a great technological improvement over shadow-dialling with a gnomon. It does not require observation of the sun's shadow at the moment

of sunrise, since it gives a clear definition of both the altitude and direction of the sun after the sun has risen. Not only is it more accurate, but it gives warnings of events and is a permanent construct that needs no realigning. 41

Below is a diagram which makes the advantages of such megalithic passages for measuring important astronomical dates visually clear.

The orientation of the stone structures in the US to the equinoxes and solstices, as well as their architectural details, are as powerful a clue that they may have been built by ancient Celts as is the presence of Ogham inscriptions, and much more difficult for scoffers to ignore.

The archaeological record makes it abundantly clear that some ancient civilizations had both a sophisticated knowledge of the celestial mechanics and a highly advanced seafaring capability, far surpassing that of the cultures that the conventional frame-work of history holds up as the most advanced cultures of the

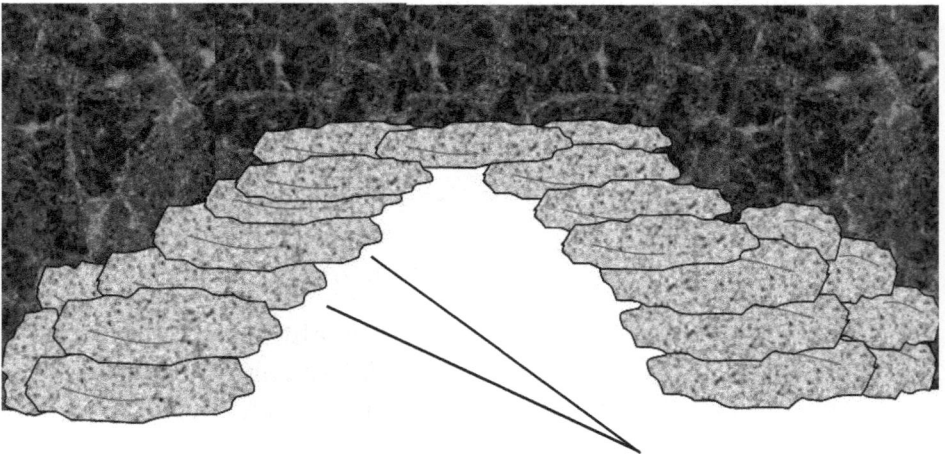

Corbels: Stones protruding beyond the vertical of the wall, used to support weight above.

Diagram of stone corbelling. A vaulted ceiling is created by the use of successive corbels until the walls are close enough to bridge with a single slab. Earth would often be packed above and around a corbelled core, as was done at Newgrange in Ireland and at Upton, Massachusetts.

ancient world. These abilities slowly dimmed and eventually were lost altogether. Contrary to the narrative children receive throughout their formal education, the earliest civilization that traversed the globe in the first millennium after the global flood was the most advanced, and those that came after retained ever hazier recollection of their technical achievements, until they lost the ability to replicate them entirely. They probably preceded the ancient Egyptians, building the great monuments of Giza and heavily influencing the civilization that followed their departure, and then at some point in time for some unknown reason left.

Curiously, Fell also describes and presents photographs of numerous stone phallic symbols discovered throughout the Americas. If the reader recalls all the way back to the account of Isis and Osiris from Plutarch, Isis searched all over Egypt for her husband's body (which had been cut in pieces by Set). She found

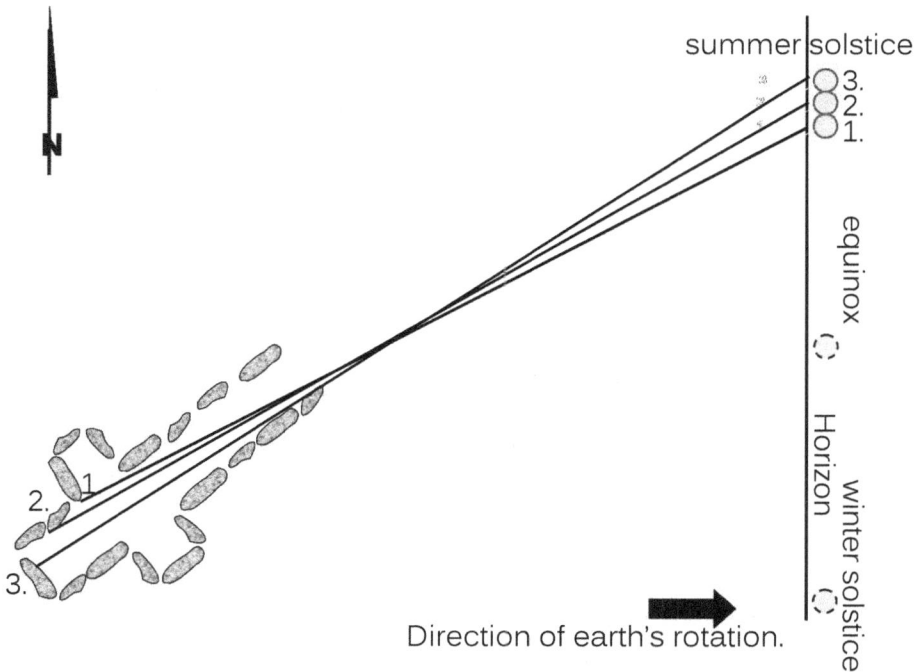

In this diagram, the sun is moving towards summer solstice rising point along the horizon, from point 1 to 2 to 3. Sunbeam will strike passage stones at 1, 2, and finally 3 on the backstone. Adapted from Brennan, *Stars and the Stones*, 42 - 44.

them all except his private member, and Plutarch relates that for this reason "Isis, in lieu of it, made its effigies, and so consecrated the phallus for which the Egyptians to this day observe a festival" (80). It appears that the evidence of this ancient practice was destroyed in much of the Old World but not in the New, and provides yet one more piece of evidence for the ancient connection of the most ancient inhabitants of Egypt with the seafarers who spread remnants of their culture throughout the world (there is plentiful evidence for the same practice in India as well).

I have taken some time to go through all the pieces of the puzzle in detail in order to pave the way for a rather remarkable theory. We have seen that Celts were apparently present in the Iberian Peninsula, writing an older form of the Ogham inscriptions that would later be most closely associated with Ireland. We have seen that these older inscriptions, like ancient Hebrew, often omitted vowels. We have seen that Phoenicians were present in ancient Iberia as well, but that the Phoenicians were not Hebrews is quite clear (the Biblical record indicates that the Canaanites were descended from Ham, and that Tarshish was the grandson of Japheth, while the Hebrews were descended from Shem).

Who, then, were the Hebrews in the Iberian Peninsula? The surprising answer is that they were the Celts! Or, to say it more precisely, the Celts were Hebrews. The clues all fit together:

- The Celts of Ireland used a form of Ogham descended from the form used in the Iberian Peninsula.

- The most ancient name for Ireland was *Ibheriu*, most likely because they came there from the Iberia, which in turn was known by that name because of the descendents of Eber who lived there.

- One of the Ogham inscriptions found in America reads: "To Baal Son of Iabagug," which is the same name as Habakkuk (the "k" sounds being voiced yields a "g" sound).

- The record of the sacred Hebrew chronicles illustrates that there was a constant tension between those who worshiped the sun-images and the "starry host" and those who cut down and burned the same and "did that which was good and right" (see for example the passages in II Chronicles 14:1-5, II Chronicles 15:8).

- This tendency to abandon the commandments and proper worship of their Creator and to worship the baals and idols is given in Scripture as the cause of the captivity of the ancient Hebrews, first the northern ten tribes and then the kingdom of Judah. The location of the "lost ten tribes" was never satisfactorily explained and was the cause of extensive speculation and analysis for centuries (although such speculation has largely been shut down by the ridicule of modern academia since the end of the nineteenth century).

- Scholars who studied Druidic religion in previous centuries were struck by similarities to Hebrew worship in what they found of Druidic worship.

- Older forms of Ogham found in Iberia and in the Americas omit vowels, just as ancient Hebrew writing did.

- "Root cellars" found in America, some of them inscribed with Ogham dedications, are clearly related to the megalithic mounds of Ireland and their astronomically-aligned passageways.

Barry Fell believes that the Phoenicians, those mighty sailors of Tarshish, transported the Celts to the Americas, where they could exploit the abundant natural resources which the men of Tarshish would then transport and trade. Celts in the Americas mined for copper and other metals, and hunted and traded furs to the Phoenician ships that would call on those shores and then transport those goods around the world.

This scenario is certainly likely, but it is also likely that the ancient Hebrews were familiar with the oceans themselves.

Many psalms contain vivid imagery of the ocean. Perhaps most significant is Psalm 107.

[21]Oh that *men* would praise the LORD *for* his goodness, and *for* his wonderful works to the children of men!

[22]And let them sacrifice the sacrifices of thanksgiving, and declare his works with rejoicing.

[23]They that go down to the sea in ships, that do business in great waters;

[24]These see the works of the LORD, and his wonders in the deep.

[25]For he commandeth, and raiseth the stormy wind, which lifteth up the waves thereof.

[26]They mount up to the heaven, they go down again to the depths: their soul is melted because of trouble.

[27]They reel to and fro, and stagger like a drunken man, and are at their wits' end.

[28]Then they cry unto the LORD in their trouble, and he bringeth them out of their distresses.

[29]He maketh the storm a calm, so that the waves thereof are still.

[30]Then are they glad because they be quiet; so he bringeth them unto their desired haven.

[31]Oh that *men* would praise the LORD *for* his goodness, and *for* his wonderful works to the children of men!

Such a description could perhaps describe a storm in the Mediterranean, but it is even more descriptive of the massive swells that prevail upon the vast open expanses of the Atlantic and the Pacific. The Psalmist is describing swells that cause the ship to mount up to the heavens and descend down into the depths, that make the sailors to stagger like a drunken man and to literally lose their wits (the Hebrew reads "all their wisdom is swallowed up").

Martin Doutré notes that a Professor Herbert Hannay, writing in 1916 (towards the end of the period in which such academic inquiry was permissible), believed that the Hebrews were at least as skilled at ocean travel as the famous Phoenicians. He said, "the greater part of the credit and renown hitherto bestowed upon the Phoenicians in connection with the maritime and colonial achievements of the Phoenician golden age should really be accorded to the Hebrews, and in particular to the tribe of Asher and southern Dan, and perhaps also, in a smaller degree, to Zebulon, Western Manassah, and Ephraim" (*European and Other Race Origins* 22, cited in Doutré 222).

Doutré notes that Deborah's victory song in the ancient Hebrew book of Judges rebukes the tribes of Dan and Asher, saying "Gilead abode beyond Jordan: and why did Dan remain in ships? Asher continued on the sea shore, and abode in his breaches [or creeks]" (Judges 5:17). He notes the prevalence of the name Dan in boat-related place-names of Europe, such as Danmark, the Dardanelles, the Danube, and in the ancient name of the Irish, the Tuatha De Danann.

Fell himself argues that the ancient Celts possessed ocean-going prowess, describing the evidence of Celtic sea power recorded by Julius Caesar, who described the high-prowed ships of the Celts and implied more than once that they were designed for the open ocean. In *De Bello Gallica*, Book III, chapter 13 Caesar says of the Celtic ships:

> The Gauls' own ships were built and rigged in a different manner from ours. They were made with much flatter bottoms, to help them to ride the shallow water caused by shoals or ebb-tides. Exceptionally high bows and sterns fitted them for use in heavy seas and violent gales, and the hulls were made entirely of oak, to enable them to stand any amount of shocks and rough usage. The cross-timbers, which consisted of beams a foot wide, were fastened with iron bolts as thick as a man's thumb. The anchors were secured with iron chains instead of ropes. They used sails made of raw hides or thin leather, either because they had no flax and were ignorant of its use, or more probably because

they thought that ordinary sails would not stand the violent storms and squalls of the Atlantic and were not suitable for such heavy vessels. In meeting them the only advantage our ships possessed was that they were faster and could be propelled by oars; in other respects the enemy's were much better adapted for sailing such treacherous and stormy waters. We could not injure them by ramming because they were so solidly built, and their height made it difficult to reach them with missiles or board them with grappling-irons. Moreover, when it began to blow hard and they were running before the wind, they weathered the storm more easily; they could bring in to shallow water with greater safety, and when left aground by the tide had nothing to fear from reefs or pointed rocks – whereas to our ships all these risks were formidable. 80.

Whether they sailed across the oceans themselves or in conjunction with the Phoenicians, it is very plausible to believe that the influence of these ancient Hebrews (Celts) or their predecessors is responsible for the circles and solar observatories found around the globe.

The theory in full, then, is that the survivors of the flood, whether by virtue of longer lifespans or by virtue of carrying advanced knowledge from the time before the flood (or both), were responsible for the Giza Pyramids and encoded in them advanced knowledge that was precious to them. They realized that the enormous cataclysm of the flood altered even the rotation of the earth, and they encoded this knowledge in allegorical form and strictly enjoined certain initiates to pass it along.

Some took these allegories of the motion of the stars and planets to be legends about actual gods and goddesses, and perhaps even some of the initiates did as well. There was clearly a tension between the worship of these celestial deities and the unseen Creator of all – a tension not only found in the Hebrew Scriptures but even among the Egyptians and the Greeks (the great Egyptian historian E.A. Wallace Budge, whom we encountered in a previous chapter, described this tension in detail and argued that there was a strange thread of monotheism running right through the apparently very polytheistic religion of Egypt, and if one reads the writing of Plato and his descriptions of the dialogues of Socrates this same strange dualism between references to multiple gods and references to a single unnamed deity is present there as well).

At some period, perhaps a thousand years after the flood, the builders of the Great Pyramid were driven to leave Egypt, and they began to migrate throughout the world, but always they preserved the encoded knowledge of the stars, the size of the earth, and the movements of the heavens wherever they went. The ancient Phoenicians, Hebrews, and the ancient Anatolian Sea Peoples who later invaded Libya and became Libyans were either their direct descendents or the recipients of some of their technologies, and some of these peoples continued to navigate the oceans for thousands of years afterwards, although the evidence appears to support the assertion that over time they became less advanced rather than more so.

When they arrived in the New World and destinations in the Pacific, they encountered peoples who had migrated previously, over the land bridges present in the first centuries after the first draining of the flood event, before the continents sank lower and the sea levels rose higher. Before the waters rose, ancient cities and megalithic structures were built in places now covered with the sea, such as Yonaguni, sites off the coast of India, the Maldives, and very likely many other sites yet to be discovered.

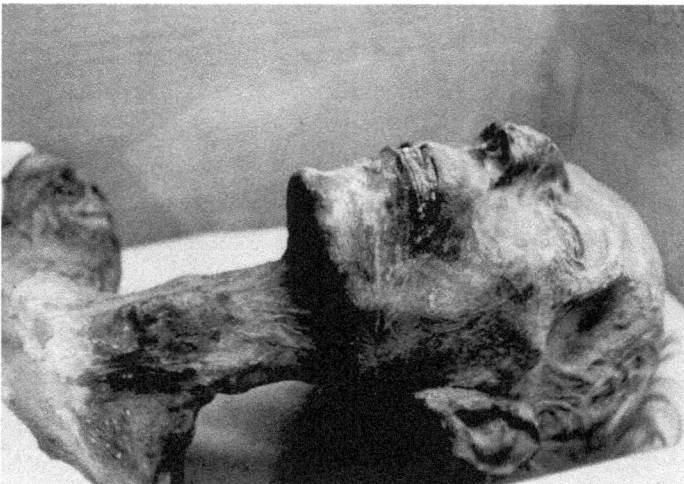

Mummy of Pharaoh Ramses II (reigned from 1279 BC to 1213 BC), showing red-gold hair. Modern microscopic inspection of the roots of the hair confirms that this reddish color was his natural hair color.

When the Celts and Phoenician-Iberians and ancient Libyans arrived in the New World and the Pacific Islands and encountered the civilizations of those who had crossed the land bridges and were there already, in many cases peaceful commerce and interaction took place, as well as assimilation, intermarriage, and two-way cultural influence. In a few cases, such as perhaps the Zuni people, their descendents remained as a linguistically-unique culture amidst the other cultures around them.

In other cases, and eventually in almost all locations in the New World, conflict erupted and one group slaughtered those who did not look like them, or drove them into centuries of hiding and furtive living.

In New Zealand, the descendents of the Celts who had sailed there in the first few centuries before Christ (prior to the complete destruction of the final Celtic seagoing vessels by Julius Caesar and the Romans) built stone circles and other astronomical markers, mined the jade "greenstone" found in the mountains of the southern island, and continued to live there largely unmolested for centuries after contact with Europe was cut off (Doutré 272-284). However, they were later driven into exile in the rugged mountains by the thirteenth-century arrival of seafaring Polynesians who were perhaps also descended from the same ancient civilization by way of the Anatolian Sea Peoples who became the ancient Libyans and settled many of the islands of the Pacific.

It is very probable that the skull of the Ruamahanga Woman belonged to a member of the descendents of these ancient Celts, who lived in the remote mountain fastness of the north island of New Zealand in the seventeenth century, some four centuries after the coming of the Maoris but before European re-settlement began in earnest in the eighteenth and nineteenth centuries. Perhaps she and her people lived in hiding along the gorges of the Waiohine or Tauhereniku, and her remains were eventually washed down into the Wairarapa to end up along the banks of the Ruamahanga.

Fell has documented connections between ancient Libyan language and writing and the language of not only the Zuni but the Polynesian languages spoken across the island chains of the Pacific. Martin Doutré has documented extensive evidence in New Zealand of ancient Celtic influence and even echoes of Egyptian art and worship in numerous articles published on his website. Examination of ancient Jomon Period pottery from Japan seems to evince similarities to ancient Libyan pottery and pottery found in the Americas in which Fell saw Libyan influence.

It is also remarkable that the Hawaiians appear to blow a conch-shell horn (or *pu*) for sacred purposes in much the same way that the ancient Hebrews use the *shofar* ram's horn, and that they use the word *aloha* in much the same way that the ancient Hebrews and Jews today use the word *shalom*, which carries a deep meaning of peace, welcome, blessing, reconciliation, respect for another being made in the image of the Creator, and a right relationship between God and man, man and man, and man and Creation.

Thor Heyerdahl (1914 – 2002), in his 1992 book *Easter Island: The Mystery Solved,* demonstrates that the features of the descendents of one group of Easter Islanders are distinctly Mediterranean in appearance, without referring to Fell's analysis suggesting ancient Libyan influence in the language of the Pacific Islands.

Other survivors of that first Giza-building ancient civilization may well have moved eastwards and influenced the ancient Hindus and Chinese, or those ancient cultures may have been descended or influenced by a different branch of the same original highly advanced survivors of the flood that built Giza.

There is also abundant written and archaeological evidence that China sent ships across the Pacific to the Americas on more than one occasion before internal struggles put an end to such voyages. There is evidence that the ancient Celts and Phoenicians mined and traded in North America, Central America, and South America, and even as settled far as Australia and New Zealand (Martin Doutré points to evidence that ancient civilizations conducted extensive mining in Australia as well).

There appears to have been very early and very extensive settlement of Central and South America by very technologically advanced and astronomically oriented settlers, who are likely responsible for many of the mysterious ancient ruins there. This conclusion is also supported by evidence that points to the same ancient people who left the region of Egypt. In addition to the North American inscriptions he describes in *America BC*, Barry Fell mentions Iberian and Libyan inscriptions found in South America, such as those in Brazil. However, some of the most incredible evidence for the extensive activity of ancient Hebrew Celts or their predecessors in Central and South America are the mummified remains of blond and red-haired people in Peru and other parts of South America documented by Martin Doutré and others.

In the vicinity of Nazca in Peru, site of the famous Nazca lines, thousands of ancient mummies have been exhumed with red, golden, or auburn hair. They are typically buried in a sitting posture, in elaborate "mummy bundles" of brightly-colored blankets and clothing, and have long braided hair – clearly not Spaniards or Europeans from after 1492 (see the work of Martin Doutré at www. celticnz.co.nz).

Mummy bundles from the region of Nazca, Peru, showing auburn and red-gold hair. Thousands of such mummies have been unearthed, and are less easily dismissed as "forgeries" than are the many tablets, steles and inscriptions in the New World.

Nor are they Incas or typical Native Americans, who have black hair and very different facial and skull features than the doliocephalic ("long-headed") mummies of Peru. Many of the mummies have wavy hair – a trait not commonly (or even uncommonly) found among Native Americans in North, Central or South America (Martin Doutré www. celticnz.co.nz).

The sheer volume of remains makes them difficult to ignore, and yet modern academia generally does ignore them, or treats these Peruvian mummies as if they are somehow clearly Inca and require no extraordinary investigation. Some apologists for the conventional storyline have tried to argue that over thousands of years, hair colors could change due to chemical processes, but scientific studies conducted by Mildred Trotter and others in the twentieth century disprove this line of argument (as Thor Heyerdahl reported in his *American Indians in the Pacific*, 1952, and as Martin Doutré has pointed out in his many online articles at www. celticnz.co.nz).

As Doutré has also explained on his website, the gene for red hair is extremely rare among the human population (between 1% and 2%), and to find it extensively in pre-Columbian mummies in South America points strongly to previously unknown European contact prior to our surviving recorded history (although not unknown in the histories of the ancient Maya, Inca and Aztec, all of whom had legends in which tall, bearded, fair-skinned and fair-haired ancients who departed their shores long before play a prominent role, as Graham Hancock has discussed quite thoroughly in *Fingerprints of the Gods*).

It is worth pointing out as well that many of the mummies of ancient Egypt, such as the mummy of Rameses II, show evidence of red or reddish-gold hair, giving further credence to the theory that the Celts and other Europeans were descendants of the original occupants of Egypt and the builders of the pre-dynastic pyramids. In the *Iliad*, Homer also describes some of the ancient Greeks (whom he refers to most frequently as Danaans and Achaians) as having red-gold hair, particularly their leader Menelaos and their champion Achilles.

There are literally thousands of mummies found in regions of South America like the mummies shown in the photographs. These are from Peru, most of them from very close to the Nazca region, within 25 miles of the mysterious Nazca lines. They are shown here because their well-preserved hair is clearly different from the Incas and other Native Americans, and provide visual confirmation of the theory we have been discussing that is immediate and visceral in a way that is even more compelling perhaps than the hundreds of inscriptions and megalithic structures described above. It is notable that, when questioned by the Spanish invaders, the current occupants of these structures all stated that there were people who preceded them who were responsible in some way for the mighty buildings that they now occupied.

The records of scholars prior to the twentieth century show a willingness to explore these sorts of clues of ancient transoceanic contact. After the twentieth century, however, such lines of inquiry were slowly choked out. Perhaps this was due to the rise of today's "political correctness" – a concept that looks at ways in which those in power (hence the term "political") keep power away from others, including through the use of language and control of what is printed or allowed in texts. Modern-day political correctness seeks to "undo" the wrongs done in the past, when those groups in power silenced those not in power, through a rigorous "re-working" of language and a hypersensitivity towards any language that could be construed as trying to impose old hegemonies again. Thus, any suggestion that Europeans might have influenced the cultures of Native Americans or Pacific Islanders is viewed as a throwback to old racist ideas and denounced in the strongest terms, whether or not such suggestions are actually motivated by racism and whether or not actual evidence leads in that direction. And yet to suppress evidence today in order to try to atone for the wrongs of past generations is neither helpful to anyone nor a remedy for racism.

Political correctness, however, is really a more recent phenomenon that is mainly a product of post-modernist and post-structuralist criticism that did not gain widespread traction until after the 1970s. In earlier decades of the twentieth century, it is likely that Darwinist assumptions muted evidence that might lead to the kinds of conclusions that were regularly reached in previous centuries. Suggestions that ancient man could have somehow crossed the oceans deliberately and regularly would fly in the face of the developing Darwinian consensus. After the end of the nineteenth century, we find such suggestions becoming rarer and rarer very quickly, until almost nobody within academia pursued such lines of argument, and those who did were outsiders and likely to remain so, subject to ridicule and generally treated with contempt by those on the inside.

After over a century of such treatment, any discussion about a worldwide flood and human survivors possessing levels of technological ability far beyond that traditionally ascribed to humans before the Egyptian, Greek and Roman civilizations (in fact, beyond that demonstrated in those civilizations) makes many people so uncomfortable that they will reject it out of hand, as a violation of their core belief system. It is likely that this rejection is built primarily upon the widely-accepted tenets of Darwinism, which the state-run educational systems in America and the UK and many other countries force-feed their charges from the earliest age until they are conditioned by decades of programming to internalize Darwinist assumptions as unassailable fact.

To return to our Scooby Doo or Sherlock Holmes analogy from the beginning of this work, we noted that in those fictional mysteries, the larger community is often convinced of some conventional explanation for the crime. The fun begins when the outsiders (Sherlock Holmes, or the team of kids with their dog Scooby) come along and challenge the preconceptions of the local police or the "experts" from Scotland Yard, and then – examining the same clues that everyone else had access to – provide a new explanation that actually matches reality much better. Darwinism is the most hardened and entrenched of all the "conventional explanations for the crime." It is the theory that the local police refuse to question – the equivalent of the assumption that the distinguished pillar of the community could not possibly be a suspect in the case.

The idea of ancient civilizations being able to traverse the globe, or even to know that it is a globe that can be traversed, and to know about the subtleties of its rotation in the heavens to a level of sophistication that far surpasses that of the average modern citizen of any advanced economy, assaults the foundations of the Darwinian model of human development to the point that some block it out entirely, and block out too all the extensive evidence that supports such a conclusion.

It is, of course, still possible to accept the theory of this book and continue to hold on to Darwinist beliefs. To do so, one must argue that the possessors of this advanced technological knowledge were themselves the products of Darwinian evolution, which took place in the distant past prior to the worldwide flood, and that all the evidence of the predecessor civilizations were wiped out by the cataclysm. However, the very idea that the earliest civilizations of which we have record were influenced by predecessors whose knowledge and capabilities surpassed their own by orders of magnitude is so contrary to the current Darwinian narrative that even hardened Darwinists will not countenance the possibility. It is probable that they realize that, should they concede that a predecessor to the Sumerians, Egyptians, Greeks and Romans knew about the size of the earth, the Golden Ratio, and the mechanics of precession to a degree far surpassing that attained by Ptolemy, and that this predecessor also knew how to navigate the globe with a level of confidence surpassing that attained by the seventeenth-century explorers, then the door to believing that the predecessor civilization was actually the product of supernatural creation would be too inviting. Thus, all the evidence must be explained away and those who continue to stubbornly insist on bringing it up must be mocked, marginalized, and called deliberate frauds or worse.

And yet even the most hardened Darwinists, in moments of unguarded contemplation, must admit that their theory is extremely untenable. We have already discussed at length that the mechanism of natural selection requires vast amounts of time, which are supplied by the uniformitarian theories. This is because the idea that mutation plus natural selection can produce the incredible variety of life – from bees that can regulate their body temperature by reversing their circulation, to humans who can compose symphonies or make three-point jump shots – is actually quite unprovable. Darwinists can only offer a lot of arguments that amount to saying "given enough time, this mechanism *could* have worked." If you examine their work closely, this is all that they actually attempt to prove with their evidence. Even though they will say in their conclusion "this is what *did* happen," all their actual evidence is of the variety to prove "this is something that *could* have happened."

Set aside for a moment that, in spite of their attempts to prove that their mechanism could work, it actually could not (in moments of honesty, even arch-Darwinists will acknowledge that their evolutionary mechanism could not work all by itself: the process which could produce self-replicating molecules such as RNA and DNA, which make the whole theory possible in the first place, cannot be explained by Darwinists, which is why when pressed on this subject Richard Dawkins had to resort to speculation that perhaps DNA was brought to earth by space aliens).

Even if we grant that the Darwinian explanation *could* have worked (which it couldn't), that is not a very good way to argue in a murder case, for example. If all the evidence points to the fact that X really committed the crime, then it would be somewhat dishonest for the defense lawyer of X to cook up a completely implausible story that Y actually did the crime, and to say "well, I know this is incredibly unlikely, but let's speculate that Y *could* have done it, since no one was there to prove that he didn't." It's one thing to make the prosecution come up with enough evidence to actually convict your client – it's a completely different thing to come up with a story that did not happen and then try to argue that this new story could actually have happened. The question is not what could have happened, but what actually did happen. The preponderance of evidence on the surface of the earth leads to the conclusion that a world-wide flood happened, and the alternative explanations (pressed into service to support another wildly speculative theory) do not fit the evidence very well at all.

The vast preponderance of evidence from man's ancient past supports the conclusion that an ancient civilization was able to cross the globe, and encode a sophisticated understanding of the earth and the heavens in amazing megalithic structures and artifacts, and that its descendents retained this globe-crossing ability as well as astronomically-inspired mythologies for centuries afterwards. The fact that these structures – even the oldest of them – are still precisely aligned to astronomical events and the cardinal directions of the earth provides powerful refutation to the reigning theory of plate tectonics.

Other evidence from mankind's past, including legends that indicate a connection between a world-wide flood and the unhinging of the celestial axis, indicate that humans actually experienced the flood first-hand, preserving this knowledge and the knowledge of the celestial mechanics in mythologies, mythologies that contain strikingly similar patterns in ancient cultures around the world, from Greece to China to the Pacific Islands.

Over time, these explanations of the celestial phenomena devolved into worship of the sun, stars and planets as gods, but this was not the original understanding but rather a perversion of the original understanding. In time, the original knowledge was lost, just as over time the descendents of those who were able to sail across the mighty ocean barriers finally lost the ability to do so, although the Celts and Iberians retained this knowledge for a significantly long period until their conquest and the destruction of their sea-power at the hands of the Romans (after which some hardy refugees probably continued still further north to continue their ship-building techniques as early pre-Viking Norsemen). Many representatives of these related ancient cultures repeatedly visited the so-called New World many centuries before Columbus. In fact, other investigators have suggested that Columbus relied upon maps and descriptions of ancient mariners who had plied the oceans millennia before.

Many of these "unconventional" conclusions have been put forth in various forms by others who have seen that the evidence simply does not support the conventional narrative of man's ancient past. The alternatives they have offered, however, often posit explanations for which there is very little evidence, such as the idea that aliens created the first humans (or brought the first humans out of beast-like existence to enlightenment), or the idea that a close encounter with Venus brought a catastrophe that wiped out a highly-advanced civilization thousands of years before recorded history.

Walt Brown's examination of the extensive geological evidence of a world-wide flood and his explanation of the mechanism by which such a flood could have come about and then receded fits the evidence we find in the mythological record and archaeological record in every way. His examination of the collision event explains why the earth experienced a massive roll, a phenomenon which would alter the heavens and which was caused by the thickening of Asia and the buckling of the massive Himalayas. It explains why tectonics is not a valid theory (which is confirmed by an examination of the continued astronomical and geodetic alignment of the Egyptian and Central American pyramids) and provides a scientific explanation for observations that tectonics vainly attempted to explain. It also provides an explanation for land bridges and the repopulation of the earth that is much more plausible than the conventional explanation.

If the geological features we see around us were caused by a huge catastrophe instead of by tens or hundreds of millions of years of gradual processes, the ages of time needed for the Darwinian theory are no longer a valid assumption. In fact, evidence suggests that this catastrophe took place only thousands of years ago, not millions. While this cuts the legs out from under the Darwinian theory, it turns out that the evidence from man's ancient past contradicts the expectations of Darwinists as well, in that man evidently went from a state of very high technological understanding (a level almost impossible to square with Darwinian assumptions, especially if demonstrated thousands of years before Greece and Rome) to a state of gradually diminishing knowledge and understanding. It is possible that even today, with our technological achievements in many areas, we still have yet to recapture knowledge that was lost, particularly when it comes to the incredible internal capacity of the human body and brain, glimpses of which are still to be seen among the lamas of Tibet, the fakirs and yogis of India, and some of the martial monks of China and Japan.

The amazing evidence presented in this chapter is not meant in any way to advance one people group over another or to suggest that some ethnicities have inherent abilities that others do not. In fact, quite to the contrary, it should reveal to us a view of man that elevates all human beings, exactly the opposite of the way that the theory of Darwinism can have the effect of bringing down our view of man to the point that he is no different from an animal. The heights which man quite clearly occupied in the first thousand years after the flood should remind us that we are all descended from people who were created, not products of evolution, and who were given amazing abilities which we have in large part forgotten. Such a message is anathema to the guardians of academic orthodoxy, and they will rush to suppress it whenever it is mentioned, and attack with the greatest venom anyone who dares to speak such words out loud.

Conclusion

So what?

We have examined a mass of evidence, from archaeology as well as mythology and even geology, and reached some fairly startling conclusions. Perhaps some readers were convinced, while others may remain skeptical. But for those who have concluded that the weight of the evidence supports the proposition that things are not exactly as the conventional guardians of academic orthodoxy would have us believe, the question still remains, "So what?"

What is the import, if we are wrong about events that took place distant millennia ago? What does it matter for our lives today?

I believe the answer may lie in another question. If it is true that, contrary to what we were taught, mankind once possessed great mathematical knowledge and understanding of the size of the earth, great technological capability to cross the seas and plumb the paths of the stars, to raise great megalithic stones and structures and build pyramids which we would be hard-pressed to duplicate today, and then slowly lost that knowledge and descended into a state that can only be described as relative civilizational ignorance, HOW DID THAT HAPPEN?

How does a civilization lose this knowledge, going backwards, slipping down a long slope for century after century? We, who occupy a point of time in which progress seems inevitable, who have lived in a generation that can look back on generations before us and trace undeniable advances in technology and in general understanding of all sorts of natural phenomena, might find it almost inconceivable that men once lived with a similar grasp of the world around them and then succeeding generations lost it or threw it away. Intellectually we might realize that this has happened before in human history, but only so long ago that we imagine those people were completely different than we are, or that we ourselves are now so different that it could never happen to us.

Yet the evidence presented in this book suggests that it did happen, and that it happened to an awful degree. Further, if we read between the lines, the evidence suggests that it did not happen simply because of laziness or the slow rolling of the ages. A civilization that built the Giza pyramids did not pull up stakes and move to Britain simply because the weather turned bad in North Africa, even though that theory has been put forth and such a change may have played some role. If we are correct in deducing that the stone circles found around the world are echoes and reminders of the Great Pyramid – Martin Doutré having argued that Stonehenge in many ways serves as a functional, half-sized, two-dimensional model of the Great Pyramid, and other monuments around the world appear to have a similar function – then we must realize that men would not simply make do with a Stonehenge after building the Great Pyramid unless they were forced to do so, most likely by violence. We can tentatively suggest that much of the dispersion can be categorized as an exile.

While we cannot be dogmatic in saying exactly what happened, our own experience as human beings in our own day and age would suggest some general categories of changes that could have brought about such violence, expulsion, and the gradual darkening of man's understanding.

Understanding and progress tends to grow when men of different nations, skin colors, and beliefs appreciate and work together with one another, and conflict and regression tend to occur when such differences begin to divide men and they start to hate and resent those of other nations, skin colors, and beliefs.

We have also seen hints of a tension between those who codified the movements of the heavens but worshiped an unseen Creator of them all, and those who somehow took these movements and turned them into a religion of worshiping the sun, moon, stars, and especially planets. One of the major insights put forth in de Santillana and von Dechend's thesis is that the planets were not named after the gods, as most of us were taught in school, but that in fact the gods were named after the planets! Somehow in the explanation of what happened to that ancient knowledge and understanding there appears to be a subtext in which men turned away from their true Creator to worship the objects of the created order – in other words, by which men inverted the natural order and placed the creation above its Creator.

In our world today, the idea that every person we meet is made in God's image and therefore worthy of protection and respect is on the wane. The idea that human life is to be protected and not taken away except to stop that human from doing violence to another is under attack from many directions. Large percentages of the world believe there are reasons to kill on the basis of different beliefs, or on the basis of grievances that members of one group have received at the hands of another group.

Many people have pointed out that ancient statuary and bas-reliefs in Central America appear to depict men of all different ethnicities, including European, Polynesian, African, and Asian, as well as American Indian or Mesoamerican. It is possible that there were periods of mutual respect and appreciation among ancient people who traded and interacted with one another, but that these later broke down. Other archaeological evidence, such as one Central American mural depicting dark-skinned warriors leading away captive light-skinned warriors, many of whom are mutilated or have their arms chopped off at the elbows, shows the kinds of conflict that signaled the end of such periods of peaceful interaction.

Another pattern of human experience with which we are all probably familiar is the deliberate suppression of facts and viewpoints that threaten the position of those in power. Ignorance and the regression of human knowledge don't "just happen" – they are often the result of the active suppression of knowledge by men with certain goals. Throughout this book, I have argued that the kind of evidence we have been examining is actively and systematically suppressed and deliberately ridiculed by those in power. In other words, one of the ingredients for the kind of backwards degradation we are talking about is already very much in place. The organs of "higher learning" in much of the developed world have been captured by those with a certain agenda, and certain lines of inquiry and investigation are off-limits. Such a situation can very well be the recipe for a plunge into many centuries that, if we could see into the future, we would label as horrible new dark ages.

Another lesson is that, if we do plunge again into another dark age, there are ways in which some knowledge can at least be preserved. It has happened before! In fact, the myths and the physical structures, texts, and artifacts we have discussed in this book are very much the voices through which those ancient ancestors speak to us across the chasms of hundreds of centuries, even across dark ages in which almost all knowledge went out. A few dedicated preservers of knowledge laboring on in hiding, such as the Irish monks in their lonely island caves, have always survived even the darkest periods.

If we remember back to the very beginning of this discussion, we contemplated the skull of the Ruamahanga Woman, who was probably the descendent of Celtic or Phoenician (or Celtic-Phoenician) settlers from ancient times, who were driven into hiding in New Zealand for centuries due to the threat of violence and extermination (Martin Doutré provides extensive evidence that ancient Europeans survived for generations in the mountain fastness of New Zealand, and to those who have read his theories, the discovery of the Ruamahanga skull in 2002 does not come as such a surprise).

We have also seen that the ancients encoded certain facts about precession, the size of the earth, and even the Golden Ratio in redundant and insistent fashion into their most enduring pyramids, megalithic structures, and even routine units of measurement and utensils. Martin Doutré has argued that this incorporation of vitally-important numbers into even mundane household and mercantile items helped ensure they would never be forgotten.

Graham Hancock and Robert Bauval have advanced the argument of Dr. Philip Morrison of MIT that magnificent monuments such as the Great Pyramid function almost as an "anti-cipher" (we in the US might call it an "anti-code") – a code designed to transmit information rather than to hide it, just as the golden plaque and record sent into space with the Voyager spacecraft in 1977 was designed by a committee to be decipherable by other advanced life forms which might encounter it (239). If this theory is correct, and it certainly seems plausible, then the discussion in this book and the works of the others cited herein attests to the success of their efforts. Even millennia later, members of a civilization with almost no memory of their existence can trace the unmistakable clues that they left and begin to piece together the outlines of the knowledge they felt was most important to transmit.

In short, it matters very much whether what we have been discussing in this book actually took place before, because it should matter very much to us to prevent it from taking place again. If it did take place, and I think the evidence we have at our disposal today indicates that it did, then we should expend further evidence to investigate how and why it took place. We should encourage scholars of all backgrounds to look into this area of human history, instead of discouraging such investigation in every way possible. And it is certain that we will not see much determined inquiry into the crucial question of how that loss took place if very few people know or believe that it actually did take place.

My own efforts to understand the evidence available are necessarily limited by my own biases and blind spots. I hope that others will also turn the lens of their own experience and education on this evidence and offer their own analysis and arguments to the conversation. Only when enough voices are raised will there be a hope of cracking the consensus and the servants of that consensus who seek to stymie all inquiry in this direction.

Finally, as argued in the last chapter, the evidence of man's ancient past should give us hope, that we are not the nearly-animal products of a long, clawing climb out of the primordial ooze, living in a dog-eat-dog world where the only two imperatives are to kill off rivals and to have sex to reproduce the species. On the contrary, we are all near brothers and sisters, descendents of the few people who were saved from the global flood, a cataclysm of unbelievable ferocity. We are all closer relatives than we have been taught in school, and we are all members of a human race that is amazingly endowed with intelligence and energy that is probably far beyond anything we can even imagine.

As the Psalmist declares, "What is man, that thou art mindful of him? and the son of man, that thou visitest him? For thou hast made him a little lower than the angels, and hast crowned him with glory and honour" (Psalm 8:5).

Thus, our shared ancient past carries both a warning and a cause for hope.

SDG

Works Cited

Allen, James P. *Ancient Egyptian Pyramid Texts*. Atlanta: Society for Biblical Literature, 2005. Print.

Bauval, Robert and Graham Hancock. *Keeper of Genesis: A Quest for the Hidden Legacy of Mankind*. London: Heineman, 1996. Print.

Benson, Miles Clifford. *Mitochondrial Genome Variation and Metabolic Traits in a Maori Community*. Masters Thesis, Victoria University of Wellington, December 2009. 31. Web. <http://researcharchive.vuw.ac.nz/bitstream/handle/10063/1224/thesis.pdf?sequence=1>

Brown, Walt. *In the Beginning: Compelling Evidence for Creation and the Flood,* 7th ed. Hong Kong: Center for Science and Creation, 2001. Print.

Brown, Walt. *In the Beginning: Compelling Evidence for Creation and the Flood,* 8th ed. Web. <www.creationscience.com>

Budge, E. A. Wallis. *Liturgy of Funerary Offerings: The Egyptian Texts with English Translations*. 1909. Web.

Caesar. *The Conquest of Gaul*. Trans. S.A. Handford. New York: Viking Penguin, 1951. Print.

Cole, J.H. *Determination of the Exact Size and Orientation of the Great Pyramid of Giza*. Cairo: Government Press, 1925.

Cremo, Michael A. and Richard L. Thompson, *Forbidden Archaeology: The Hidden History of the Human Race*. Badger, CA: BBT Science Books, 1996.

Cumont, Franz. *Mysteries of Mithra*. Trans. Thomas J. McCormack. Chicago: Open Court, 1903. Web. <http://openlibrary.org/books/OL7029629M/The_mysteries_of_Mithra> image of mithraeum found at <http://www.archive.org/stream/mysteriesofmythr00cumouoft#page/n73/mode/2up>

De Santillana, Giorgio and Hertha von Dechend. *Hamlet's Mill: An essay on myth and the frame of time*. Boston: Nonpareil, 1969.

Doczi, György. *The Power of Limits: Proportional Harmonies in Nature, Art, and Architecture*. Boulder, CO: Shambhala, 1981.

Dodwell, *The Obliquity of the Ecliptic: Ancient, mediaeval, and modern observations on the obliquity of the Ecliptic, measuring the inclination of the earth's axis, in ancient times and up to the present*, 1963. Web. <www.setterfield.org/Dodwell_Manuscript_1.html> 16 Oct 2010.

Doutré, Martin. *Ancient Celtic New Zealand*. Auckland: Dé Danaan, 1999. print.

Doutré, Martin. *Ancient Celtic New Zealand*. Web. <http://www.celticnz.co.nz/>

Fell, Barry. America *BC: Ancient Settlers in the New World*. New York: Demeter, 1976. print.

Fletcher, Alice Cunningham. *The Omaha Tribe*. In *27th Annual Report of the Bureau of American Ethnology to the Secretary of the Smithsonian Institution, 1905-1906*. 1911. Web.

Frankel, Henry. "The Continental Drift Debate." in Hugo Tristram Engelhardt and Arthur L. Caplan *Scientific Controversies: Case Solutions in the Resolution and Closure of Disputes in Science and Technology*. Cambridge: Cambridge UP, 1987. 203-248. Print.

Frazer, James George. *Folk-lore in the Old Testament: Studies in Comparative Religion, Legend and Lore*. London: Macmillan, 1918. Web.

Frazer, James George. *Golden Bough: A Study in Comparative Religion.* London: Macmillan, 1890. Web.

George, A. R. *The Babylonian Gilgamesh epic: introduction, critical edition, and cuneiform texts,* vol. 1. NY: Oxford UP, 2003. Print.

Hancock, Graham. *Fingerprints of the Gods.* New York: Crown, 1995. Print.

Hancock, Graham. *Underworld: the Mysterious Origins of Civilization.* New York: Three Rivers, 2002. Print.

Hapgood, Charles H. *Maps of the Ancient Sea Kings: Evidence of Advanced Civilization in the Ice Age.* Kempton, IL: Adventures Unlimited, 1966. Print.

Heeres, J. E. *English Translation of the Journal of Abel Janszoon Tasman.* Project Gutenberg of Australia, 2006. Web. <http://gutenberg.net.au/ebooks06/0600571h.html>

Herodotus. *Histories.* Aubrey de Selincourt, trans. Suffolk: Penguin, 1972. Print.

Homer. *Odyssey.* Trans. Robert Fagles. New York: Penguin, 1996. Print.

Keeler, Clyde E. *Secrets of the Cuna Earthmother: A Comparative Study of Ancient Religions.* New York: Exposition, 1960. Print.

Krupp, Edwin C. *Echoes of the Ancient Skies: the Astronomies of Lost Civilizations.* NY: Oxford UP, 1986. Print.

Mair, A.W. (trans) and G.R. Mair (trans). *Callimachus: Hymns and Epigrams, Lycophon and Aratus.* 2nd Ed. Loeb Classical Library 129. London, 1921. Print.

McEvilley, Thomas. *The Shape of Ancient Thought: Comparative Studies in Greek and Indian Philosophies,* New York: Alworth Press, 2002. Print.

Mitchell, Stephen. *Gilgamesh: A New English Version.* NY: Free Press, 2004. Print.

Neugebauer, Otto. *History of Ancient Mathematical Astronomy.* Providence: Springer-Verlag, 1975. Print.

Piazzi Smyth, Charles. *Our Inheritance in the Great Pyramid, New and Enlarged Edition.* London: Isbister, 1874. Web.

Plutarch. *Isis and Osiris.* in *Plutarch's Morals,* vol 4. William Watson Goodwin, trans. Boston: Little, Brown, 1874. Web.

Plutarch. *Life of Pompey.* in *Plutarch's Lives.* Bernadotte Perrin, trans. Loeb Classical Library, vol 5. 1917. Web.

Poignant, Roslyn. *Oceanic Mythology: the myths of Polynesia, Micronesia, Melanesia, and Australia.* London: Hamlyn, 1967. Print.

Rey, H. A. *The Stars: A New Way to See Them. Enlarged World-Wide Edition.* Boston: Houghton Mifflin, 1966. Print.

Sellers, Jane B. *Death of Gods in Ancient Egypt: A Study of the Threshold of Myth and the Frame of Time.* Rev. Ed. Lexington, KY: Jane B. Sellers, 2002.

Stewart, Matt. "New Theory in Skull Mystery." *Wairarapa Times-Age.* 21 May 2009. Web. <http://www.times-age.co.nz/local/news/new-theory-in-skull-mystery/3798060/>

Swerdlow, N. M. *Review of* Origins of the Mithraic Mysteries: Cosmology and Salvation in the Ancient World. *Classical Philology,* vol 86. number 1. January 1991. 48-63. < http://people.sc.fsu.edu/~dduke/lectures/swerdlow-mithras.pdf>

Tolkien, J.R.R. *Two Towers.* New York: Ballantine, 1965. Print.

Tomkins, Peter. *Mysteries of the Mexican Pyramids.* New York: Harper & Row, 1976. Print.

Ulansey, David. *Origins of the Mithraic Mysteries: Cosmology and Salvation in the Ancient World.* New York: Oxford UP, 1989. Print.

Warder, Anthony Kennedy. *Indian Kavya Literature, Vol 6.* 1992. Web.

Waters, Frank. *Book of the Hopi.* New York: Penguin, 1967. Print.

Yarwood, Vaughn. "Written in Blood." *New Zealand Geographic.* Number 96, March/April 2009. 36-47. Print.

www.ingramcontent.com/pod-product-compliance
Lightning Source LLC
Chambersburg PA
CBHW021044090426
42738CB00006B/183